Cichlids of Lake Tanganyika

Part 1

The species of the genus Tropheus

Peter Schupke

Die folgenden Bildautoren trugen zur inhaltlichen Gestaltung dieses Buches bei:
The following photographers contributed to this volume:

F. Teigler	F. Schäfer
H.-J. Mayland	H. J. Richter
E. Schraml	H. W. Dieckhoff
Sch. Nakano	A. Konings
L. Seegers	P. Louisy
B. Bender	K. A. Ackerbauer

Ferner dankt der Autor in besonderer Weise den folgenden Personen für Tiere, die Vermittlung von Bildmaterial und/oder vielfältige sonstige Unterstützung.
Special thanks to the following persons for fishes and/or other help:

Helmut Löfflad, Marc Danhieux, Sven Danhieux, Thomas Lepel, Thorsten Reuter und Hans-Georg Emmerich für die von ihnen zur Verfügung gestellten Fische, sowie für die Zeit und Geduld, die sie mir bei meinen vielen Fragen entgegen gebracht haben

Wolfgang Staeck, Hans-Joachim Herrmann, Ad Konings, Patrick Louisy, Laif DeMason, Lothar Seegers, Horst W. Dieckhoff und Hans-Joachim Mayland für die von ihnen zur Verfügung gestellten Bilder.

Im besonderen Siegmund Janicki und Helmut Löfflad, die es ermöglicht hatten, neue Tropheus-Populationen im ehemaligen Zaire unter schwierigen Bedingungen zu fangen und zu exportieren, bzw. zu importieren.

Seinen Bekannten in der Aquaristik für die gemeinsamen Diskussionen und Anregungen in den ganzen Jahren.

Seiner Frau, die den Großteil des Textes eintippte und besonders seinen beiden Söhnen für ihr Verständnis, dass nicht sie, sondern er die meiste Zeit vor dem Bildschirm verbrachte.

Buntbarsche des Tanganjikasees, Teil 1:

Die Arten der Gattung *Tropheus* / Peter Schupke

[Übers. Dt.-Engl.: Mary Bailey)

Rodgau: A.C.S., 2003

(Aqualog)

Deutsche Ausgabe: ISBN 3-936027- 37-4

English edition: ISBN 3-936027- 38-2

Liebigstr. 1, D-63110 Rodgau / Germany

http://www.aqualog.de

Fax: +49 (0) 6106 644 692

Text und fachliche Bearbeitung
Peter Schupke

Übersetzungen
Deutsch-Englisch: Mary Bailey

Redaktion
Frank Schäfer

Titelgestaltung
Verlag A.C.S.

Druck, Satz, Verarbeitung
Lithos: Verlag A.C.S.
Bildbearbeitung: Frank Schäfer

Druck: Westermann Druck, Zwickau
Gedruckt auf EURO ART glänzend,
100% chlorfrei von PWA, umweltfreundlich.

PRINTED IN GERMANY

Umschlagfotos / *Photographs on cover*

Front cover:
Tropheus sp. "Kirschfleck"

Back cover:
Tropheus sp. "Rutunga", *T.* sp. "Ikola", *T.* sp. "Mabilibili", *T.* sp. "Kachese"

Alle Fotos / *all photographs*: P. Schupke

Content

* by Christian Sturmbauer

** by Frank Schäfer

The lake

Lake Tanganyika is one of the largest bodies of fresh water on our Earth. It is not just its enormous size, but also its age, that makes it so interesting to aquarists and scientists alike. According to the most recent research the lake has been in existence for about 12 million years – an extraordinarily long period of time, in which a multitude of life forms have been able to evolve. On the basis of the large number of species of bivalves, snails, and sponges, it was once assumed that the origins of the lake lay in the sea; but more recent research has demonstrated that without doubt Lake Tanganyika, which lies in the Great Rift of East Africa, gradually filled with water as the result of the drifting apart of two continental plates, with the Malagarasi and other rivers making a significant contribution. Because the Malagarasi and its numerous affluents drain a very large area, some of the cichlids of the lake probably derived from fluviatile species. According to the latest research, however, it can also be hypothesised that a number of fluviatile cichlids have their origins in the lake. To cite just one example, the rheophilic *Orthochromis*, which SEEGERS & DEVOS (1998) discovered in Tanzania, exhibit wide similarities to the goby cichlids of Lake Tanganyika. Another rheophilic cichlid, Schwetzochromis neodon, which Bleher and I collected in 1991, is endemic to Lake Fwa and the River Fwa in the Central Congo. This river is ultimately linked with the Congo, which, prior to the formation of Lake Tanganyika, may have been joined to the Malagarasi. The fluviatile cichlids from the Congo and Tanzania, and the goby cichlids of Lake Tanganyika, are very similar in a number of respects

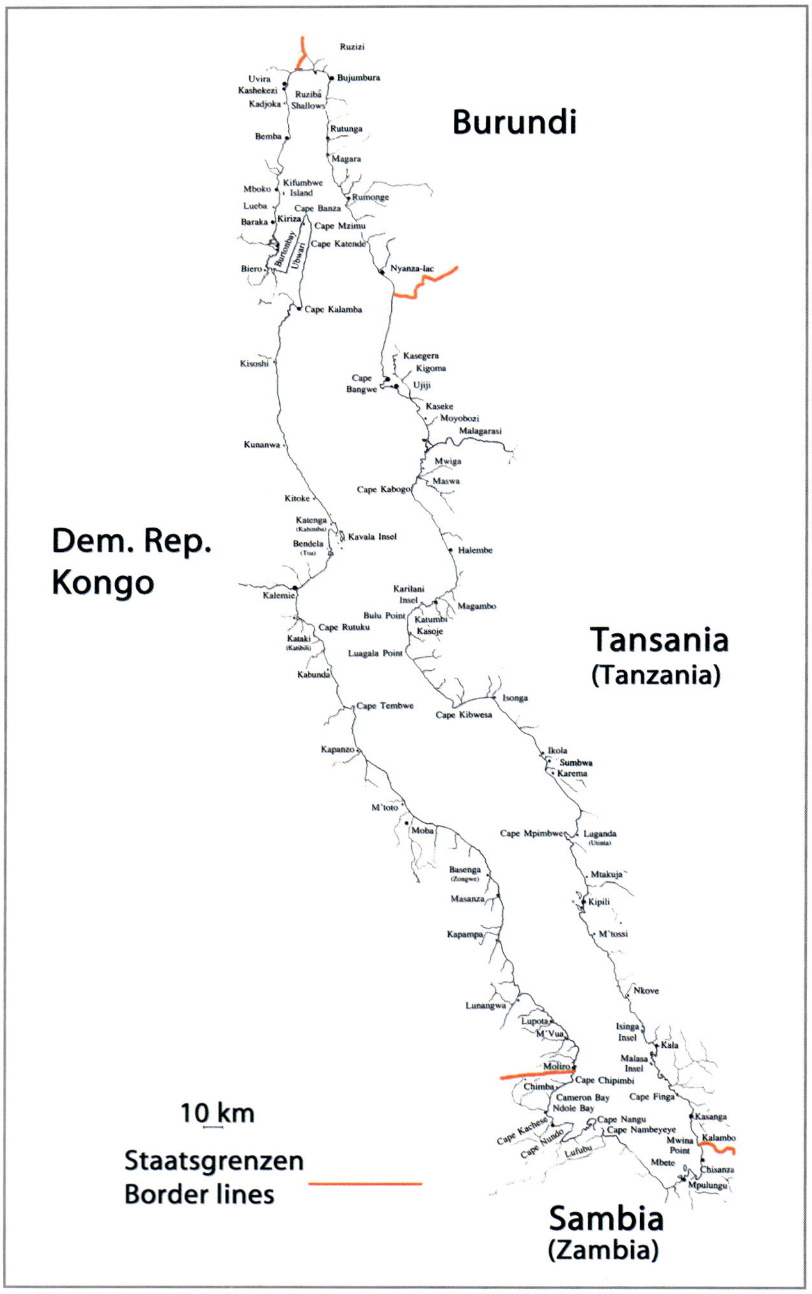

Orthochromis luichensis photo: L. Seegers

Schwetzochromis neodon photo: P. Schupke

Eretmodus cyanostictus photo: P. Schupke

The water

But let us return to the water of the largest Rift Valley lake. Except where it is affected by rivers flowing into the lake, the water may be regarded as being medium hard, and, with a pH of 8 to 9, very alkaline. Its conductivity (more than 600 micro-siemens) is relatively high, and this can be explained by the constant influx of minerals from the rivers. Under the higher pressure in the lake, certain mineral salts precipitate out and form a thick whitish layer in several metres of depth, coating the entire rocky littoral. Magnesium is one of the chief causes of hardness in the lake. The water in much of southern Germany and many other areas worldwide has a high conductivity and is thus well-suited for the maintenance of the fishes of Lake Tanganyika. This high conductivity is based largely on calcium and magnesium salts. Unfortunately it is impossible to say whether, or to what extent, the high calcium content of such water is tolerated by the cichlids of Lake Tanganyika. The magnesium content of the lake is relatively high (144 mg/l) and the calcium content correspondingly low (30 mg/l). The high pH value of the water indicates that little free CO_2 is present. This should be borne in mind, as in the aquarium high population densities and dirty filters can cause a serious rise in carbon dioxide levels and the pH may drop to levels that the fishes can no longer tolerate. The oxygen content of the upper levels of the lake is very high due to good circulation. Only at deeper levels does the oxygen content decline dramatically; at 200 -250 metres of depth the oxygen content is so low that only anaerobic life can exist. Sometimes it happens that, because of the topography of the lake bed, water from the depths containing sulphur dioxide rises towards the surface during severe storms, and may reach the levels inhabited by deeper-living fishes (Herrmann, 1997). The lake then presents a very sorry picture, with tens of thousands of fishes washed up on the shore. According to Herrmann, these dead cichlids consist largely of deeper-living species such as *Cyphotilapia, Cyprichromis, Xenotilapia* and *Bathybates* species. The temperature of the surface waters of the lake fluctuates seasonally to a considerable extent, and these variations cannot simply be ignored. In February and March (autumn in the southern hemisphere) I recorded a surface water temperature of 28 - 30°C, but a number of publications give the temperature in August (winter) as about 24°C. These temperature variations probably have a not insignificant effect on the metabolic processes of the fishes. During the rainy season the sun shines for only a few hours during the already short days (about 12 hours). During this period Aufwuchs production is not very high, and I have been unable to find any noteworthy Aufwuchs in the many rocky zones. The rocks were in fact very clean. At certain times of the year vast plankton blooms occur; some species feed exclusively, others partially, and a few seasonally on this resource. Many Aufwuchs-feeders avoid rocky biotopes with a heavy coating of detritus, as in such areas there is no guarantee of adequate algal production because of the constant deposition of sediment.

Clouds over Ndole Bay, Zambia. photo: F. Schäfer

The rocky littoral close to the shore is, with one exception, the most important habitat for *Tropheus* species - only here is an adequate food supply for these fishes guaranteed. The rocky habitat may be divided into a number of zones. Pebbly beaches offer numerous hiding-places and thus harbour a large number of very young cichlids of many species. The water is oxygen-rich and warm. Because of the shallow depth the Aufwuchs grows well and provides sufficient food for the young fishes. But there is also a higher predation pressure in this zone, and in consequence the species that live here are generally cryptically coloured. This changes in species that abandon this biotope in the course of their development. The growing fishes migrate into the rocky littoral and seek out crevices appropriate to their size among the rocks. Half-grown, occasionally even full-grown, Tropheus have a preference for football-sized boulders, but the majority of adult Aufwuchs-feeders require aggregations of larger rocks, on the one hand to satisfy their need for security, and on the other because this habitat provides the large rocky surfaces necessary to feed numerous large fishes. The large rocks, several metres square, which in addition lie close together and contain numerous caves, are the home of the largest of the Aufwuchs-feeders of the genus *Petrochromis*. They have an enormous requirement for food that can be satisfied only here, and. defend their rocks with great vigour. In this part of the rocky littoral, which represents a not inconsiderable portion of the shoreline of Lake Tanganyika, many other fishes of a huge variety of genera find refuge in time of danger. In these biotopes we find all sorts of Aufwuchs-feeders as well as piscivores that feed on species living in or over the rocky littoral. One example is *Cyphotilapia frontosa*, which feeds on, inter alia, fishes which it captures in the early morning twilight period.

There is, however, yet another biotope where at first glance we might not expect *Tropheus* and *Petrochromis* to be at home. This is the sandy coast. But just one look beneath the water's surface provides the explanation. Baked sand, so-called sandstone, riddled with innumerable holes and crevices, offers these fishes protection and a certain amount of food. The population density in this marginal biotope appears correspondingly sparse. This biotope seems not particularly suited to *Tropheus*, as their population density is not very high. *Petrochromis* are likewise sparse in their occurrence. This region is in fact rather more suitable for the genera *Simochromis* and *Lobochilotes*, which are correspondingly numerous here. The areas of sand between the sandstone rocks also provide pure sand-dwellers with a home. In such mixed biotopes fishes completely different in their requirements and appearance are encountered together. The constantly murky water is not conducive to extensive algae production. The coarse and not particularly surface of the sandstone does not provide optimal living conditions for algae.

Ndole Bay in western Zambia contains a biotope of this type. Even so, these sandstone rocks provide a wonderful opportunity for travelling from one rocky biotope to another. In consequence the boundaries between one colour population and the next are likewise not clear-cut, and sometimes even fluid, where sandstone biotopes of this type separate the populations, whereas large open expanses of sand present an almost insuperable barrier for *Tropheus*.

In Ndole Bay the bottom is mainly sand, but there are also large expanses of sandstone here and there. photo: P. Schupke

Fish communities

The rocky littoral is very rich in species, offering them simultaneously a high degree of security and food. With a few exceptions, the majority of species live here permanently or part of the time. The permanent residents, of course, include almost all the members of the Tropheini tribe, which as well as Tropheus includes the genera *Petrochromis, Simochromis, Pseudosimochromis,* and *Lobochilotes* (which will be dealt with in the next volume in this series).

A few species of the tribe Ectodini (which I will likewise cover later) are also found in the rocky littoral. Thus we regularly encounter the genera *Ophthalmotilapia, Xenotilapia, Asprotilapia, Cunningtonia*, and, on the periphery, *Cyathopharynx, Lestradea*, and *Callochromis*. Not forgetting the Lamprologini (an enormous number of species), including *Altolamprologus, Neolamprologus, Chalinochromis*, and *Julidochromis*. Because of the extensive food supply, many piscivorous cichlids are likewise regularly encountered, examples being *Boulengerochromis, Bathybathes, Ctenochromis*, large lamprologines, spiny eels of the genus *Afromastacembelus*, and the highly venomous water cobra. The larger piscivores have the unusual habit of normally hunting in small troops. I was able to observe this spectacle every day of my visit to the lake. My first experience of this kind plunged me into a state of simultaneous fear and curiosity. I was sitting in the water near the shore, watching all the hustle and bustle in the rocky littoral. In a matter of seconds all the fishes stopped what they were doing and disappeared in panic into the nearby caves and crevices. Something must have caused the fishes to take flight. Alarmed, I surveyed the surrounding area. Then a small group of elongate fishes came into view, at first just shadows, but rapidly coming closer. They swam very close to my spot, and a brief glimpse was enough to identify them as the cause of the panic - they were half-grown *Boulengerochromis microlepis*, hunting for food. Then, as quickly as they had fled at the approach of these predators, the boldest rock-dwellers were again leaving their hiding-places to devote themselves to the continuing search for food.

Always present, albeit not in large numbers, scale-eaters hover at some interest from the bottom, forever looking for a victim to target. With a little patience these menaces can be observed snatching their food. It seemed to me that the victims were not particularly concerned; after a successful attack by a scale-eater they immediately resumed their daily round.

The littoral is also very species- rich at depths of about 3 - 5 m. Where it falls away to greater depths, fishes that normally live much deeper can regularly be seen. These spots are often used by ornamental fish collectors, in order to catch species that would otherwise be out of range. As well as juvenile Aufwuchs-feeders, the young of the larger piscivores – such as *Bathybathes* – are also found in the littoral close to the shore; and here too are often found species such as *Haplotaxodon*, which normally occur in deeper water and further from the shore.

The food supply in the littoral

The entire rocky littoral is overgrown with a coating of algae of variable thickness, which, depending on the area, may be covered with a layer of sediment. Many species make use of this Aufwuchs. It is, quite simply, the staple food of Aufwuchs-feeders. Other species use it some of the time to supplement their menu. One species of algae, which is related to the Cyanobacteria, lives free-floating in the water. This is the food mainly of open-water species such as *Cyprichromis*, sometimes *Ophthalmotilapia* and others. These free-floating algae are of little importance for the genera of the Tropheini tribe.

In the course of their evolution the Aufwuchs-feeders have adapted very well to this poorly-digestible food. Their digestive systems, and in particular the intestine, are specially designed to process this food. This requires a long intestine and a special digestive secretion that breaks down the food so it can be used by the fish. Aufwuchs-feeders spend almost the entire day collecting food, and are never sated like, for example, a piscivore after a heavy meal. We should not forget this when maintaining these fishes, as otherwise we run the risk of their overeating. This is the start of a long road of suffering for the fish (and the aquarist), culminating in a corpse. With a little common sense this fate can be avoided. As already mentioned, Aufwuchs may contain a very large amount of sediment composed of various ingredients. A large part consists of fine sand. This sand is in turn composed of numerous minerals and is undoubtedly essential for good digestion. After a water change we can take handfuls of the sand that should always be used as the aquarium substrate, and sprinkle it over the rocks in the tank. In this way the fishes are obliged to graze sand along with the algae on the rocks.

Cape Chaitika; a scale-eater is swimming in the center of the picture.
photo: P. Schupke

The aquarium water

Cichlids are in general very tolerant as regards many water parameters. This is evidenced by their huge adaptability all around the globe. We find them in extremely soft water, for example in the tributaries of the Amazon in South America, and also in the extremely hard water of the African soda lakes.

In aquarium terms, this adaptability means that they can be kept in most types of mains water. But aquarists with greater ambitions, for example breeding, cannot avoid creating water conditions for their fishes that approximate very closely to those of their native waters.

In the specific case of Lake Tanganyika cichlids this means a pH above 8, although in a few cases a pH of 6.5 is adequate. But a pH of 8 or higher is better, and should occur automatically in hard water with strong aeration (to drive off CO_2). Many plants do not thrive very well in this alkaline environment. In such water biogenic calcification may be a problem. In practical terms this means that a fine, sometimes very hard, film of calcium carbonate is deposited everywhere, especially in places particularly devoid of CO_2 - mainly heaters, but also many other solid objects in the tank. To prevent this it generally sufficient to remove the airstone (or other diffuser). These serve mainly to drive off CO_2 and introduce only negligible amounts of oxygen into the aquarium; the latter is diffused far better via movement of the water's surface.

For longer term maintenance, a conductivity of about 100 µS/cm to 1000 µS/cm, sometimes more, has proven acceptable. Naturally the ideal is for the conductivity to match that of the water in Lake Tanganyika, which is about 600 µS/cm.

The high mineral content of tap water in many areas makes it ideal for the maintenance of East African cichlids. It also ensures a good pH buffering capacity, which is important in heavily populated tanks. For this reason soft water from the tap should be hardened to an appropriate level.

In the case of extremely soft water it is perhaps a good idea to reconsider, and instead keep cichlids or other fishes from soft water, for example from South or Central America or from West Africa.

Much has been written about filtration, which remains a topic of considerable debate. A lot of the advice given is good, but equally some is ridiculous. I propose to discuss here only a few fundamental points regarding filtration that I consider important.

A filter must be able to perform two important functions:

1. Collecting visible dirt such as droppings and mulm. This mechanical part of the filter must be cleaned as often as necessary (in heavily populated tanks = daily).

2. A filter must provide a substrate for additional nitrifying bacteria. This biological part of the filter may also be regarded as a "bacteria farm" or a bacterial culture. These two main functions are often only poorly fulfilled by commercially available filters. It is true that nowadays there are multi-chambered filters with a preliminary, easy-to clean pre-filter, but these have only a small market share and are correspondingly high-priced. The universally familiar canister filters, which stand outside the aquarium and are connected to it by hoses, can perform only one of our two functions. However, if they are fitted with an appropriate pre-filter, then they too can perform the complete purification and conditioning of the water, although the very high throughput is not particularly conducive to efficient bacterial activity. Bacteria may colonise such filters initially, but after only a short time the media become dirty and clogged; the bacteria then do not obtain sufficient oxygen and their activity becomes limited. Moreover, such filters are tricky and time-consuming to clean. The first prerequisite of a good mechanical filter is that it should be quick and easy to clean. There is a tendency to put off cleaning an inconvenient filter. The filter medium for a mechanical or pre-filter can be filter foam or floss. It should always be washed in hot water in order to kill off any potential pathogens (which often find conditions in such filters to their liking). Bacterial or biological filters come in all shapes and sizes.

It matters not whether it is a very expensive trickle filter (with the disadvantage of a high evaporation rate), a built-in internal filter, an external filter, or a seed tray such as I use above the aquaria in my cellar. All must function adequately using the biological principle. All biological filters are containers in which bacteria are cultivated. The composition of the filter media in such bio-filters must also serve two purposes. Firstly it must offer the bacteria as many surfaces as possible to colonise. I personally prefer to fill my seed trays with ordinary filter wool and every couple of months I squeeze it out, very briefly but thoroughly, in the trays in order to remove old dead bacteria and mulm. Interference with the media should be kept to a minimum as otherwise too many live bacteria will be destroyed.

The epitome of convenience is the blue foam mat, generally 10 cm thick, which extends across the width of the tank and is fixed to the front and rear glasses at one end of the tank or the other. Unfortunately it has one insoluble drawback: it must break down solid wastes with the aid of bacteria. In the course of time mulm accumulates in the filter block and the result is often fishes with an increased respiratory rate and whose condition is not always optimal, evidencing a degree of ignorance which, unfortunately, is widespread in the aquarium hobby. Bio-filters of any kind should not be allowed to draw in solid wastes. They should be powered by a low-turnover powerhead or by air. The throughput should be no more than about 25% of the aquarium volume per hour, so that the water is in contact with the bacteria for sufficient time. The water parameters in an active bio-filter should be checked occasionally, while it is running. The pH of the out-flowing water should be about the same as that entering the filter. The oxygen content should approximate to that of the aquarium water. If it is lower, then the throughput should be increased. If that doesn't help, then the filter is no longer functioning satisfactorily and should be partially cleaned or some of the media replaced. The media in a bio-filter should never be completely changed all at once, as otherwise the majority of the bacteria will be lost. New tanks can be rapidly matured so that the water is suitable for fishes by rinsing the media from an established filter in the water of the newly set-up tank. The mechanical filtration will clear the water overnight, and the tank will now have a small bacterial population which will rapidly proliferate. If after every water change the bacterial population is supplemented using a bacterial starter (this is strictly necessary only in heavily populated tanks with too small a bio-filter) then the result will be crystal-clear water, fishes in superlative condition, and excellent resistance to disease. The regular use of a good bacterial starter can even replace the biological filter in a normally to heavily populated aquarium. When the bacteria are used up, the water will rapidly be affected and may become cloudy. This shortage of bacteria will be evident from the fishes: initially they will react with an increased respiratory rate, and later by appearing unwell. In the latter case, repeated re-dosing is necessary. But this should be the exception rather that the rule.

Water changes

There are many ways and means of avoiding changing water. Nitrate can be removed using ion-exchange resins, in order to put off such a change; but there is no real substitute for changing water. The only real justification for ion-exchange is to purify seriously polluted mains water in order to be able to keep fishes at all. Nitrate-removing resins may work very well, but they have the disadvantage that what they remove from the water by way of nitrate, they add in the form of chloride (especially in the case of ion-exchangers for hard water). Or else they remove not just nitrate but also the calcium carbonate in the water – which is equally undesirable. A weekly water change remains the best way of removing nitrate and refreshing the water.

I was once told that the regular water changes (as a rule weekly, and – depending on population density – in exceptional cases up to half of aquarium volume) to which my fishes were subjected, would significantly reduce their tolerance of heavily-polluted aquarium water. The source of this reproach was a dealer who refused to take any more stock from me unless I performed no water changes for several weeks on tanks containing the fishes he had ordered. This just goes to show the sort of things our fishes have to put up with.

Every tank has its own individual population, and thus its own optimal volume of water to be changed, which should not be exceeded. As already mentioned, a balanced relationship between volume, population, and the optimal maximum water change will manifest as very clear water, healthy fishes, and an aquarium with excellent function. The entire system will function beautifully until such time as we add new fishes or the existing occupants have grown and upset the environmental equilibrium through the increase in metabolic wastes. One might think the answer would be simply to change more water, but this doesn't work. While every tank is, of course, different, there are limits beyond which one simply cannot go. The relationship between volume of water and fish mass functions only up to a certain point. That is the key to the whole thing. And it is something that we can learn only by experience.

Even *Tropheus* and *Petrochromis* – wild-caught or tank-bred – should rarely fall ill in a properly matured tank. The aquarium should, however, always have previously contained fishes as otherwise the bacteria will have no "food" and when the new fishes are added the nitrifying capacity of the bacteria will only subsequently attain optimal efficiency. An increase in nitrite will be unavoidable. Nitrite is badly tolerated by all fishes, and even the smallest amounts inevitably lead to death.

Temperature

In my opinion, the question of the temperature at which Tanganyikan cichlids should be kept is invariably given insufficient attention. Lake Tanganyika is subject to seasonal temperature changes, and we should not forget this when keeping its fishes. 24°C is not suitable for these fishes on a permanent basis. The cool water appears to make them lethargic. Their metabolism is on the back-burner. And this raises a further avoidable problem with these fishes. If the diet is too rich then they can develop digestive problems. Their systems cannot cope – or only inadequately – with the highly nutritious food that they are generally offered. Bacteria and other organisms may colonise the gut and have a negative effect on the food pulp, perhaps even causing the production of toxic metabolic products through their activity, and thereby causing additional harm to the fishes. Fishes are not warm-blooded like us humans, and for better or worse they are dependent on ambient temperature. This probably plays a subordinate role in the lake, as the food supply is completely different to our "growth foods" and high value frozen foods. Koi enthusiasts have long been aware of this, and the composition and quantity of the food offered to their fishes is based on the water temperature. At lower temperatures, for example, they are given food with a lower protein content than normal. During the East African winter, the water of Lake Tanganyika has a temperature of about 24°C and hence is significantly cooler than in summer, when the water temperature measures from 27 to 30°C. This has an effect on the metabolism of the fishes. We will thus be creating very natural conditions if in summer we set the temperature higher, and in winter lower.

At the same time we must remember to feed more sparingly, and on food with a high roughage content, at the lower temperature.

Population density

There are as many different opinions regarding aquarium size as there are on filtration. Essentially, aquarists' tanks, no matter how large they may be, are mere teacups in comparison with the lake. Aquaria with a volume of 2 cubic metres are undoubtedly large, but again pitifully small if we compare them with an actual underwater biotope. But, as we have already seen, cichlids seem to be highly adaptable, and hence relatively small tanks are suitable for a group of *Tropheus*. Unfortunately it is often the case that we are obliged to cram them into tanks that are too small to even halfway permit the development of a peaceful social structure within the group. To give a detailed example: a group of 14 *Tropheus* "tail-stripe", which had lived together for more than a year with their previous owner, and knew and tolerated one another well, had to make do with a tank 100 x 50 x 50 cm because no larger aquarium was free. The two males occupied the left- and right-hand halves of the aquarium, and as a result only the front glass area was left for the females – this was very good for observing them, but their swimming space was strictly limited because of attacks by the males. Obviously this couldn't be allowed to continue. I may have a large selection of aquaria, but there were fishes in every tank. Hence, as is so often the case, a major reshuffle had to be organised. Eventually the 14 *Tropheus* ended up in a 250 x 80 x 60 cm aquarium. "Now", I thought, "they have enough room". But far from it. The attacks by the males were now even more violent than before and I had to act more quickly than before. And now good advice proved worth its weight in gold. Yet again several communities had to be moved around, which, of course, was not always to their liking. At long last a tank measuring 150 x 60 x 50 cm became the permanent home of the *Tropheus*, and in this they lived for a long time without problems. The secret in this case was that the group was too large for the 250 litre tank, too small for the 1200 litre tank, but just right for the 450 litre aquarium. The relationship between number of fishes and tank size is, as is well known, subject to strict rules which it is advisable to obey. As usual, there are exceptions that prove the rule. Because *Tropheus* are relatively intelligent fishes and highly individualistic in their behaviour, quite simply the only solution is to experiment until you get it right. One thing is certain, come what may, in order to keep these fishes for a long time without losses we must maintain them at higher population densities than under natural conditions. Or else we can keep them in pairs, as I myself had to 20 years ago through lack of funds, although the experiments I had, of necessity, to conduct at that time did take place in large aquaria with a length of 1.5 m upwards. This method is, of course, suitable only for the aquarist who doesn't need or wish to make money from his fishes. The females of pairs kept in this way regularly produce fry, which cannot be counted on in larger groups because high level of stress. In addition, both partners remain almost permanently in good colour, which again is not always the case in large groups. My personal observations in Lake Tanganyika have repeatedly indicated that *Tropheus* normally maintain a certain "personal distance" from their neighbours. I have never seen serious quarrels in the lake, although the population density of *Tropheus* in suitable biotopes can be every high. I have seen the same sort of thing in European brook trout, *Salmo trutta* f. *fario*, which likewise need to keep their distance from their neighbours in the stream. If they are obliged to live at closer quarters then there are constant quarrels ending in the death of the weaker fish. If they are kept together in a shoal then there are rarely any battles, but their natural behaviour is significantly distorted.

Large water-worn boulders are very suitable decor for *Tropheus* aquaria, and minimise the likelihood of injury. photo: Archiv A.C.S.

Decorating the aquarium

The decoration of the aquarium is generally a compromise between personal taste and practical considerations. Lots of rocky structures with plenty of caves and hiding-places often looks very good, but can very rapidly lead to problems, as I will explain in greater detail below. The fishes will certainly like plenty of caves and hiding-places. This is fine for the majority of Tanganyika cichlids, which come from areas of the lake where there are innumerable caves and other retreats. In the lake the fishes require a shelter from predators and a safe spawning site where they can breed completely disturbed. And in Lake Tanganyika there are several hiding-places for each fish. We can try to simulate this in the aquarium as well, only the available bottom area is rarely sufficient. In such restricted quarters there will sooner or later be very serious problems, and the victims are invariably the smaller and weaker members of such a crowded aquarium community. The territories lie far too close together, and fishes without territory almost invariably find themselves on someone else's patch. Luckily such a community will also work without hiding-places. Normally it is quite sufficient to break up the line of sight in several places using large rocks or other decor objects. The higher the better. This has the great advantage that the

occupants are not permanently in sight of one another. One of my own tanks is a good example, a set-up in which I have kept *Tropheus* species for a long time without major problems: on a bottom surface of 150 x 60 cm there are 3 - 4 football-sized rocks, positioned at roughly regular intervals along the tank. These are sited in the rear half of the tank in order to create enough space at the front for individuals without territory. These "open territories" are probably not sufficiently attractive to be defended on a permanent basis.
Males do in fact hold territories in this underwater landscape, only this territoriality is of short duration and, in my experience, occurs only at breeding time. Except when spawning the males are almost invariably tolerant of territorial infringements. Many of the members of a *Tropheus* group will voluntarily remain in the untenanted front half of the tank and are only occasionally kept in order by the "top fish". They are thus subject to much less stress. For the same reason, if Tropheus are kept in very small tanks – this should never be more than a purely temporary measure - then all decorative material should be removed, and this may create peace in the tank for a while.

Many an aquarist's dream: a splendid group of *Tropheus* sp. "Kiriza" (Pop. 1.8) grazing on a rock-face. photo: E. Schraml

Decor materials

Depending on taste and finance, almost any type of decor can be used. Water-worn stones from rivers are particularly good because they are very natural, but they must accordingly be very well cleaned in order to avoid introducing disease. As a matter of principle the rocks should have rounded edges. Under no circumstances should sharp-edged material such as lava or broken flints be used because of the extraordinarily high risk of injury. Bricks – the type with holes for keying mortar – are very practical, albeit not exactly biotope-correct. They are good for providing hiding-places for smaller fishes and can serve as refuges for brooding females. I personally prefer light-coloured bricks, because algae grow on them very rapidly, and all Aufwuchs-feeders enjoy scraping off this growth. Algal growth tends to be significantly lower on darker stones. I advise against wood of all types, because in the course of time it may occasionally release substances over which we have no control. In individual cases, however, the addition of wood to Lake Tanganyika aquaria can produce positive results. In the lake I have several times seen Lamprologus species that have found desirable residences in empty drinks cans – they were obviously perfectly at home there, as they were even spawning. But, of course, cans should never be placed in the aquarium because of the danger of metal poisoning in the small volume of water.

A few large rocks are more decorative than lots of small ones.

Plants

Lake Tanganyika is no plant paradise. A possible reason may well be the high alkalinity and low CO_2 content of the lake water. Hardly any plants enjoy a pH between 8.5 and 9. The plants that are found in the lake are mainly species of the genera *Vallisneria*, *Ceratophyllum*, and *Potamogeton*. These are found chiefly in calm bays, harbours, etc, but not on the rocky littoral. For this reason decided at a very early stage to dispense with plants in my tanks. Earlier attempts to establish plants in my tanks had, moreover, failed. The pH can, of course, be adjusted by population density, aeration, and/or the use of chemicals, but I regard providing the optimal environment for the fishes as preferable to manipulation of this kind. But at lower pH and oxygen levels there is no bar to planting. This is, of course, not natural, but it can produce an eye-catching aquarium for the domestic setting. Fine-leaved plants should be avoided, as they are very likely to be regarded as food by the fishes.

Aquatic plants are uncommon in Lake Tanganyika, and fishes that prefer vegetation are instead found in the reed zone. Hornwort (*Ceratophyllum*) is well suited for a Lake Tanganyika biotope aquarium, as on the one hand it is found in the lake, and on the other it is completely undemanding as regards water chemistry. photo: F. Schäfer; drawing: Tropica, Danmark

Substrate

The substrate should not be of coarse gravel as this will collect large amounts of detritus and start to putrify. For practical reasons a substrate of fine sand, not too white in colour, is recommended, using a layer 0.5-1 cm deep. Detritus will remain on the surface of this type of substrate and be carried to the filter by the current. Moreover, the fishes will show better colour over a light-coloured, but not white, substrate.

In aquaria with overall dark-coloured decor many *Tropheus* will merge with their surroundings and look more or less like lumps of coal, which is, of course, not what we want. White sand very rapidly becomes coated in algae and then no longer looks good. At each water change I run my fingers through the sand so that it never becomes compacted. The reduction in the bacterial flora resulting from the water change is somewhat compensated by the bacteria living in the sand. An alternative option is to leave the substrate undisturbed to provide the fishes with an additional surface from which to graze algae. Given suitable lighting, then provided the water is not too polluted the quantity and quality (respectively) of the algae will be such as to create a very attractive carpet which will be consumed with enjoyment. The "scars" left by feeding, and even the various techniques used in grazing, will be evident. It should, of course, go without saying that there is no place for algae-eating catfishes in such aquaria, as these can render the entire tank "squeaky clean" overnight, leaving nothing for our cichlids. Aquaria without substrate are far from attractive and thus rarely seen in practice.

Maintenance with other fishes

All large Aufwuchs-feeders, such as *Petrochromis* and *Tropheus* species, tend to dominate any aquarium community in which they are kept in large numbers, and this does not leave very much free space for fishes with different behavioural patterns. The result is, in effect, closed communities in which other fishes find no long-term peace. At times I have *Tropheus* variants in almost all my tanks, in order to observe and photograph them. At such times my numerous featherfin mouthbrooders and sand-dwellers cease breeding and I can tell that they are clearly stressed. Among *Tropheus* there are individuals that will attack the finnage of other fishes. They are particularly fond of swimming stealthily up behind featherfins with their long ventrals in order to relieve them of a piece of fin. I have sometimes even had the impression that in *Tropheus* this is sheer wickedness. I have never seen anything as cunning and underhand in any other fish. *Petrochromis* have a similar but even more dangerous method of "seeing off" other fishes. They will scrape the protective mucus coating clean off their bodies. Continuous attack of this kind can rapidly lead to the demise of the victim. I at first thought that this behaviour resulted from some unknown dietary deficiency, until I realised that only particular fishes were targeted. My *Petrochromis* had a preference for females of *Ophthalmotilapia* species. At first I was totally puzzled, but later, thinking about the fish community in Lake Tanganyika again, a possible explanation for this bad behaviour occurred to me. The females of the species in question move through the rocky littoral in large shoals, visiting the tops of rocks large and small in order to feed. The larger rocks are generally occupied in each case by a *Petrochromis*, which is quick to defend its food supply with all its might. Competition for food can cause problems in the aquarium. *Tropheus* are problem-free in this respect, but almost all *Petrochromis* species react very strongly to tankmates and in many cases can drive the aquarist to distraction. I too have made repeated attempts, but in all but one case these proved doomed to failure. It may be that the aggressiveness of these species will gradually diminish in the course of a few generations of captive breeding. The Eretmodini tribe, which comprises the genera *Spathodus, Eretmodus*, and *Tanganicodus*, are virtually ignored by both *Tropheus* and *Petrochromis* species. They are suitable for any tank. Unfortunately these goby cichlids – in particular the *Spathodus* species – can be very quarrelsome among themselves. However if they are kept in pairs in tanks upwards of 1.5 metres long, there should be no problems with them. Unfortunately the choice of tankmates for *Tropheus, Petrochromis, Simochromis* and others of this group is not at all easy, not only as a function of their lifestyle but also because dietary considerations can be problematical. However, I will go into this in greater detail in the chapter on feeding.

Julidochromis and members of the *Lamprologus* group, which feed on small food items, are very suitable. They are well able to withstand the "bully boy" behaviour of *Tropheus* and *Petrochromis*, provided the tank is large enough. They generally form pairs and defend their territories with vigour. Almost all species that are able to take care of themselves, and can be fed on different food, are suitable. Specialist feeders such as Altolamprologus compressiceps, inter alia, should not be kept in such a community as their crustacean foods are unsuitable for Aufwuchs-feeders. Peaceful fishes will obtain too little food and become emaciated. Fishes of Ectodini and *Cyprichromis* groups can be added in large aquaria with a volume of several cubic metres. In some cases these combinations may also work in smaller tanks, as long as these are not too heavily populated. If the aquarist is not worried about being biotope-correct, then cichlids from both of the other large East African lakes can be added. In terms of character, the mbuna of Lake Malawi and many haplochromine species from Lake Victoria are very suitable. They are all belligerent bullies that get on well with others of the same kind. A carefully chosen mixed diet will feed them all satisfactorily, and they can be kept together for a long time. For reasons of space I myself have had to create such colourful communities in some of my aquaria; I encountered no major problems requiring me separate them again. A truly colourful community, rivalling many a marine aquarium.

Julidochromis species (this is a juvenile *J. marlieri*) are eminently suitable tankmates for *Tropheus*. photo: Sch. Nakano

Maintenance of a single species

Keeping *Tropheus* together warrants a chapter of its own. There are many possible ways of doing this, and what works for aquarist Smith may prove a disaster for aquarist Miller. In theory, the simplest and best method is to buy tank-breds and put them together in the tank they are to occupy as adults, so that they will have plenty of time to get used to one another. This method, however, requires a degree of patience on the part of the aquarist – but then again, in the long term patience is an indispensable prerequisite of the aquarium hobby

I have myself grown on many *Tropheus* and a number of *Petrochromis* species. I have regularly parted with adult fishes (bred by others and wild-caught) because I had enough of their offspring to create a larger group. On the basis of experience I have a mistrust of out-sourced fishes. Often they start to fall sick after many months, but this never happens with my home-bred stock; moreover it is not uncommon for peace to reign only when such "stranger" fishes are sold on. The diseases they carry can be transmitted to other fish species. Importers need to have a huge amount of experience and "feel" to acclimatise wild-caught fishes properly and appropriately for the species. But even then problems can occur. In such cases affected specimens must be sent to a reputable specialist in fish diseases (ask your vet to recommend one) for an accurate diagnosis, before the fishes can be treated appropriately. Many medications are no longer effective because they have all too often been used incorrectly or following an incorrect diagnosis. Long-term usage can also lead to pathogens developing resistance to one active ingredient or another. If, however, for whatever reason, you nevertheless still wish to buy well-grown fishes, then the new community should be established in a tank of their own. The situation is quite different with juveniles or half-grown specimens: the more the merrier. As already mentioned, *Tropheus, Petrochromis,* and some other species must be kept in large groups in order to subdue the "individuality" of each fish. Of course, cost may decide whether you buy 10 or 20 *Tropheus* sp. "Kirschfleck" ("Cherry Spot"). Experience has shown that 10 adult *Tropheus* are too few for a 1.5 m tank, and there is no doubt that 15 or more would be better. I have reservations about endorsing the recommendation that n excess of males will stabilise the group. In the course of time a dominant male will generally establish himself and regard all conspecifics that are unwilling to spawn as irritants. Their sex is of little relevance. If several males are able to occupy territories in a suitably large aquarium, then this will be an even greater catastrophe for the rest of the occupants, as now they will generally have nowhere to escape to. In such cases, hiding behind the internal filter may be their only salvation.

If unacquainted fishes are to be put together, then allow them 2-3 weeks to settle down, as it often takes this amount of time for them to sort themselves out and get to know one another. Serious quarrelling often stops after this period of time.

White mouths in *Tropheus* are a sign of intense fighting for rank.
photo: Archiv A.C.S.

Maintaining several colour populations together

The housing together of several colour populations must be properly thought through in advance, in order to avoid hybridisation. In this regard, we can make the basic assumption that neighbouring populations of the same lineage will generally show a tendency to cross. Exactly the same applies to populations whose distributions are in fact separated by a long stretch of coast, but again belong to the same lineage. The following case, which happened to me many years ago, will serve to illustrate that populations do cross. The forms concerned were *Tropheus* sp. "Ikola" ("Kaiser") and *Tropheus* sp. "Bulu Point" ("Kirschfleck"). The offspring resulting from this adulterous escapade had the external appearance of *Tropheus* sp. "Ikola". Only on closer examination did I notice the light red at the base of the scales in the yellow band in these fishes. Because, then as now, such crosses were regarded as undesirable, I left this unintentional experiment at that. The numerous colour populations certainly offer an interesting area of research in this respect. A further experiment, this time with *Tropheus* sp. "Moliro" and a single female of *Tropheus* sp. "Kabeyeye" ("yellow rainbow") lasted for more than a year and demonstrated that these populations are apparently neither visually similar nor genetically close – no hybridisation took place. Testing and documenting such combinations is just one of many necessary researches which aquarists as well as scientists can perform, in order to learn more about the evolution of these large species groups. *Tropheus* sp. "Bulu point" and T. sp. "Green", which was imported in small numbers several years ago from the then Zaire, have proved to be easy to cross.

It is normally possible to keep together widely-separated populations such as *Tropheus* sp. from the north and *Tropheus kasabae* (rainbow varieties) from the south. As usual, there are exceptions that prove the rule, so every aquarist must make up his own mind whether or not to risk it. There should be no problems with unwanted crosses between the various *Tropheus* species such as *T. polli, T. brichardi, T. moorii, T.*

kasabae, *T.* sp. (various populations to the north and south of Kalemie, Congo), and *T. duboisi*. Over the years I have kept more than 30 different populations together and not one of these combinations showed any inclination to hybridise. But I must add that I kept together only *Tropheus* populations of different lineages. The different populations, except where they belong to the same lineage, no longer "understand" each other's behavioural cues sufficiently - often they quite simply ignore each other. It is, of course a prerequisite that each population has a choice of suitable partners.

The following are a few examples that I have tried myself:

T. moorii (black) with *T. kasabae* (rainbow variants)
T. moorii with *T.* sp. "Kipili" (Tanzania)
T. sp. "Moba" (*polli*-like, Congo) with all other *Tropheus* species
T. duboisi with all other *Tropheus* species (this mix is reputed to lead to hybridisation, but I have never seen this and hence cannot confirm)

The following are examples of mixes that can be housed together with great circumspection:

- all populations of a closely related lineage, eg *T. moorii* (Kiriza) with *T. moorii* (Rutunga) or *Tropheus* sp. ("head-stripe", Zambia-Congo) with *T. kasabae* (rainbow variants, Zambia)
- *T.* sp. "Ikola" with *T.* sp. "Bulu Point".

The populations of a single lineage should not be housed together, eg the *Tropheus* of the "southern rainbow" or the "red head-stripe", etc.

Ro reduce it to a rule of thumb: the localities of the individual populations should be as far apart as possible. There is nothing fundamentally wrong with crosses, as they too are fishes worth keeping. But they must be declared to be such. In most cases the results of such crosses are likely to be uninteresting in their coloration. They are usually either intermediates, ie they inherit half the colour of each parent, or one coloration is dominant, ie it suppresses that of the other form almost completely (eg in the cross between *T.* sp. "Ikola" and *T.* sp. "Bulu Point"). I find the selective breeding of individual populations far more interesting. Thus, for example, we have obtained *T.* sp. "Ikola" with a very broad band or two narrower ones ("double spot"), *T. moorii* (Bemba) with a broad band, *T. duboisi* with broader bands. Cleanly marked *T. moorii* ("Brabant"), or *T. moorii* "Caramba" selected for increased red, would be well worthwhile. Or we can concentrate on maintaining the natural coloration and vitality of the various populations, which is itself difficult enough. This type of selective breeding, whatever its specific object, requires ability on the part of the breeder and - and this seems very important to me - continuity. Unfortunately many *Tropheus* are purchased by beginners or new businesses that have inadequate knowledge of the "essentials" of fishkeeping or have little interest in them, wanting nothing more than a tank full of attractive fishes.

Now that we have seen that it is generally possible to keep different populations together, let us take a look at another problem. All *Tropheus*, like many other cichlids, behave according to the motto, "Might is right". A very self-confident male of one colour form can generally dominate and terrorise even a very distantly related *Tropheus*. As mentioned at the start of this section, this un-enjoyable and in the long term nerve-wracking situation can be avoided by putting all the future occupants of a tank together in their new home at about the same size. But even then changes may be necessary later on. *T. moorii* "Moliro" and *T.* sp. "Canary Cheek" of the same size get along very well even if they are first put together as adults. On the other hand, the "Moliros" will not accept an introduced female *T. kasabae* ("Nangu"). Quite simply, we may need to experiment with different combinations. Sometimes we will be in luck, and it will work right away.

The situation is otherwise with some *Petrochromis* species and colour forms. A few tricks are necessary to keep large individuals together. The reasons for this are rooted in their ecology. As already mentioned, almost all species of this genus defend fixed territories on a permanent or temporary basis, territories that must be extensive enough to provide an adequate food supply for the fish. They rigorously drive away any fish that comes near. Small fishes of other tribes are often tolerated, eg *Lamprologus, Julidochromis*, and *Chalinochromis* species.
In the lake *Petrochromis fasciolatus* with its highly cryptic colour pattern sometimes occurs in large shoals that move through the rocky littoral. They are easier to keep together in the aquarium than are their cousins from other species. Some species can be readily be kept in groups until they attain sexual maturity. This was the case with my 10-strong group of *Petrochromis* sp. "Gold" from the border zone between the Congo and Zambia. They moved around the tank in a shoal and showed no inclination to thuggery up to a certain size. But overall, the genus *Petrochromis* is unsuitable for the aquarium. But if an enthusiast has nevertheless decided on one of the very colourful races, then there are ways and means of keeping their high aggression within bounds. Anyone with enough money can obtain a very large group of perhaps 20 to 30 individuals and keep them – as already described for *Tropheus* species, in tanks that are neither too large not too small. The best thing is to experiment. This "needs must" solution is certainly the easiest, but it is not satisfactory in the long term as the fishes are packed in practically like sardines and the stress will do them no good. An often-used solution involves keeping a group consisting of one male and several females, provided the females tolerate one another. If the females quarrel among themselves, then as long as they are different sizes we can provide them with suitable caves which the larger individuals cannot enter. A further solution involves using a grid divider, through which the larger fishes cannot pass. If we use the various-size caves method, then the smallest fish and its cave should be put in first and allowed several days to become accustomed to its surroundings. Then the next smallest fish should be added, along with a larger cave. And so on. These and other such contrivances are often the only way to obtain young and reduce the aggression somewhat over successive generations.

Keeping *Simochromis* species together presents similar difficulties to those with the species already discussed. *Simochromis curvifrons* is generally regarded as impossible to keep, even though this species has a number of very attractive colour varieties (underwater photos). The same applies to *Lobochilotes labiatus*, which are likewise difficult to keep together.

Provided the possibility of hybridisation and their aggression are taken into account, the cichlids of the Tropheini group can very readily be kept together without too much squabbling. An aquarium containing a community of these colourful cichlids with their varied behaviour can present a splendid spectacle, that often reminds me of my experiences in the natural habitat on the rocky littoral of Lake Tanganyika.

Feeding in the aquarium

With a few exceptions, all the members of the Tropheini group inhabit the rocky littoral where they occupy a variety of feeding niches according to their individual requirements. Aufwuchs, in all its many manifestations, represents the main component of their diet. The feeding apparatus of the fishes is just as varied as the algae. There are, inter alia, pickers, raspers, pluckers, scrapers, etc. They spend most of the day consuming food, as the sometimes sediment-coated Aufwuchs is not very nutritious and is available in only small portions. In order to obtain sufficient energy they need to consume large quantities, and a lot of energy is used up straightaway in the feeding process. Tropheus, for example, usually stop feeding only to breed. Thereafter they rasp away vigorously again, making up for lost time. I have never seen energy-consuming territorial disputes during my underwater observations. Depending on the area, the Aufwuchs is covered to a greater or lesser extent with sediment, which makes it even harder to digest. This food requires a special digestive apparatus – all species have in common a very long slender intestine, specially designed to digest mainly vegetable food. This also means a high roughage component, which the gut is likewise designed to handle. The food pap is soft and very malleable, which helps keep the digestive function healthy. And this is the source of the main problems with the maintenance of specialised Aufwuchs-feeders. A diet that includes too much protein and fat, accompanied by only a very small amount of roughage, can completely disrupt the gut. It may perhaps, with difficulty, manage to digest food of animal origin, but the roughage component is too small. The pap is then firm and inflexible, it remains in the convolutions of the gut and causes a blockage that usually results in the death of the fish. The same problem occurs in the human digestive system: too little roughage results in a hard pap and leads to all sorts of diseases, while a higher roughage component means no such problems. Similar problems arise with manufactured foods such as flake, sticks, granules, etc. These supposedly valuable foods are compounded from a number of raw materials, but the end product does not contain the correct balance of ingredients such as is found in algae or insect larvae. The digestive system of the fishes has evolved to handle such natural foods and is designed to deal with specific percentages of their ingredients. Now, you might well assume that these materials would also be contained in manufactured food. But how many ingredients do we actually know about? How many enzymes, for example, which act as catalysts in the body, are still functional in these artificial foods? Natural foods contain numerous substances whose effects are still unknown, and are sometimes lumped together as vitamins. But please don't get the impression that I am totally against manufactured foods. I know many aquarists who, for reasons of time, etc, use them exclusively. Flake, granules, and sticks all have a value as supplementary foods. An argument commonly put forward is that frozen foods are appreciably more expensive than manufactured foods. But if we keep records for a few years then a quite different picture emerges. The expense of maintaining fishes mainly on manufactured foods is considerable. Natural foods cause only minimal pollution of the water, and hence the amount of water to be changed will be far less – a factor to be taken into consideration in this day and age. The health of the fishes should not have to be expensively assured with medications and equipment. Some aquarists may thus wonder, just how cheaply can fishes be kept! Many years ago I repeatedly tried to explain to the regularly visiting representative of a large German food manufacturer just how valuable a manufactured food with an appreciably higher roughage content would be. At that time the available foods were practically all growth foods. Unfortunately this problem wasn't recognised at that time, and isn't taken seriously enough even today. And while the vegetable foods from all companies now contain a valuable component, Spirulina algae, but the roughage element is still far too small, or in other words, the food is still too nutritious. I cannot concur with arguments such as "The market for such food is too limited". Frozen food mixtures have been available for a long time, but we should check out such foods more closely before using them. If we are not satisfied with such a food, there is the option of making our own food mix which will enable us to feed our fishes healthily for many years at no great expense. And in this case we should also try unconventional foods. Thus *Petrochromis* species can be offered potato, which they will rasp away happily once they become accustomed to the idea. Equally suitable are cucumber, peas, carrots, flaked oats, inter alia. A daily menu for our Aufwuchs-feeders can look something like the following:

First there should be green food in the form of spinach (available frozen, but unfortunately blanched, in small handy packs), dandelion, lettuce, and the like, preferably frozen or fresh if the fishes will take them (this usually requires an acclimatisation period of several weeks). Overcooking should be avoided as heating damages important elements in these foods. Dandelion is a very valuable green food. With a little preparation it can form the main component of a frozen food mix for Aufwuchs-feeders. It takes little effort to shred and freeze it. Green food of this type is not always accepted by the fishes immediately, but once they have become accustomed to it, they will devour it greedily. Because, given a free choice of foods, fishes will usually eat green food last (if at all), then they should be fed this food first, followed by more popular items such as Daphnia and Cyclops (see also pages 20-21, disease in the aquarium), and then finally those foods that they would otherwise eat first, eg adult Artemia and small krill. In this way the amounts of the different foods can be reduced. Artemia and krill should be given in only small quantities, and the same applies to Mysis and river shrimp. A *Tropheus* or *Petrochromis* should never receive more than one or two small pieces of the latter food (or similar) per meal. These small amounts of animal protein can be readily handled by the Aufwuchs-feeder gut. But never feed larger quantities of these foods or blockages may occur. The digestive system of a fish can "break down" in the longer term (several months or years) if unsuitable foods are fed beyond strict limits. Thus I myself had a few specimens of *Tropheus* und *Petrochromis*, who had killed their partners, in an extra tank which contained mainly individuals of haplochromine species from Lake Victoria. Their food contained far more animal protein. As well as the obligatory spinach, they received a lot of Mysis and mosquito larvae of all types. They were allowed sufficient time to adapt to this diet. It is known that in captivity the digestive systems of some mammals and birds will adapt to unnatural foods when they have been captive-bred for generations: their intestines become shorter. Such animals are then unable to live in the wild! Feeding is always something of a minefield – if we feed a lot of green food, then the fishes are not always optimally conditioned for breeding, and if we feed too much animal protein, then the fishes soon develop serious digestive problems. If manufactured food is fed as well, then it is beneficial to use only those types with the highest green food component. The algae in our aquaria are another natural source of green food, although they will grow in adequate quantity and quality only in moderately populated tanks with good lighting. *Petrochromis* are of limited suitability for community set-ups, as these highly-specialised "lawn-mowers" will strip an aquarium of all its algae within a few hours or days. "Home-grown" algae-covered stones are one answer, but only for the aquarist with just a few tanks.

Luckily for us, most fishes are very adaptable. Wee must give them a chance to get used to our artificial foods, which should be as varied as possible in order to provide sufficient essential dietary elements, whether known or unknown. In practice this translates as follows:

- Always feed just small quantities several times a day (if you go out to work, use an automatic feeder, which should be refilled every day as some elements in manufactured foods deteriorate very rapidly).
- If possible, always feed the fishes yourself, as other members of the family are often too generous with the food.
- First of all feed a food rich in roughage and with a high green component (eg spinach), and only then a staple food such as flake with a high Spirulina component, if necessary preparing and freezing such a food yourself.
- Frozen foods, such as insect larvae, plankton, etc should be given only as a treat, and in very small amounts.
- No food should reach the bottom; better to feed too little and then a bit more
- Try unconventional foods such as various vegetables.

Don't lose patience during the acclimatisation period. It may be several weeks before the fishes will accept the unnatural food. And hungry fishes are less likely to be shy toward their owner. Later on, spinach in particular will be devoured greedily. A precept of which I am daily reminded, and which I always nevertheless find rather difficult to put into practice, is, only a hungry fish remains a healthy fish. Aufwuchs-feeders should never be satisfied.

This *Petrochromis* sp. "Texas" impressively illustrates the mouth of an Aufwuchs-feeder
photo: H. J. Mayland

Acclimatisation: basics

The acclimatisation of new fishes (and not just wild-caught stock, which the beginner should generally avoid) requires a number of measures, which may not guarantee continued survival in the strange environment, but at least offers a greater chance of keeping losses to a minimum. Wild-caught fishes will not have eaten for some time, because of the collection procedure, the need to purge their guts in the exporters holding tanks prior to transport, the journey itself, and the first few days at the importer's. Their digestive "engines" will be at a stand-still. If we ensure that they are slowly started up again using food suited to their systems, then the most difficult part of their acclimatisation (and that where most losses occur) will be behind us. But to achieve this, various points, already in part discussed, must be borne in mind.

A view of the holding area of the company A.C. Aquaculture, in Zambia. photo: F. Schäfer

Selecting stock at the dealer's

The subsequent health of the fishes depends on the professionalism of the exporter and importer. If wild-caught fishes are not properly cared for after collection, not despatched as soon as possible, and, finally, the importer fails to cater to their requirements, then it may happen that entire consignments are literally ruined. The fishes will die, even months later, with a huge variety of symptoms and under an equally wide variety of maintenance conditions in the tanks of aquarists and professional breeders. In such cases, one or more links in the chain described above has been deficient. Study prospective purchases for a long time from a distance. Pay attention to their general demeanour. Slightly frayed fins will generally heal again rapidly. Cloudy eyes likewise clear up rapidly if the water is of suitable quality, but such cloudiness may be the start of a difficult to cure bacterial infection of the lens, which can often lead to the loss of the eye. The eyes should not be protruding from their sockets – an unmistakeable sign of bad water – but under improved conditions the eyes will return to their normal position with no permanent damage. A warning: the eyes may also protrude in cases of dropsy, although in such cases the body of the fish will be severely distended. This disease is incurable. Likewise, if the eyes are sunken, leave the fish where it is. The region between the eyes and the dorsal fin should be rounded, never angular. A crease across the head is a sure sign that the fish has suffered irreversible damage and will not live much longer. The belly should be concave if the fish has only recently arrived, as it will not have received anything to eat for a long time. Ask the importer what food he has given the fishes, and obtain a small supply of this food to use for the first few days. During the settling-in period the fishes should not be subjected to another change of diet. Individuals that stay away from the main group, and those that behave abnormally, are best not purchased.

The *Bora Kula* was chartered to transport certain fishes from the east coast to the west, for export. photo: F. Schäfer

Feeding during acclimatisation

Many fishes are imported to Europe in autumn, winter, and spring, and often become severely chilled in transit, so that their metabolisms are on the back-burner. Accordingly we should never take them home and feed them with highly nutritious foods. In general it is better to leave acclimatisation to the importer, but if you nevertheless want to take on this job yourself to make sure, then put the fishes in a tank with plenty of algae and raise the temperature over a period of about one to two weeks. In such a tank they will be able to start feeding again in accordance with their individual needs. Once you notice that they are rasping algae, check that their faeces are of normal consistency as well. If so, then they can be given the tiniest quantities of their intended future diet – just enough to be eaten within a few seconds if they accept it. In order to give any that are slower than the rest a chance, then for the time being it is permissible to leave uneaten food in the tank for a short time, but if it is not eaten within half an hour then it must always be siphoned out, as at tropical aquarium temperatures it will soon start to decay. Rotten or rotting food is often a contributory factor in fish losses. Thus the new fishes will slowly but surely become used to our artificial foods. During this period the temperature should be no lower than 26° and can even go as high as 28° C. The temperature should be raised in stages over several days. Ask the dealer at what temperature the fishes have been kept. In relatively warm water the metabolic rate will increase proportionately, as evidenced by healthy appetites among the fishes. The amount of food offered should be increased only slowly, so that the fishes' systems can adjust to the new food. Don't be taken in by the eager fishes crowding around the front glass. They will not starve, even if their bellies are rather concave. During the settling-in phase, newly hatched Artemia are an excellent supplementary food. Along with the algae growing in the aquarium this is the best food for the initial stage of acclimatisation. The Artemia should be from a hygienic culture, which means, in practice, starting a new culture every day in a clean bottle. Cultures that have been running for several days will contain innumerable germs that will enter the aquarium along with the shrimps and can cause digestive problems in Aufwuchs-feeders.

Filtration during feeding

It is a good idea to use a time-switch to switch off the filter at the start of feeding. The timer should be set so that it automatically switches the filter back on after about half an hour or a little less, as otherwise it is easy to forget. The reason for this action is obvious: the filter should not be allowed to suck in food during feeding, as this will pollute the water unnecessarily, and we want to maintain water that is as free of germs as possible. The majority of Aufwuchs-feeders, especially when stressed, react very badly to high numbers of germs, and decaying manufactured food will quickly lead to a high population of germs (within a few hours). Given good biological filtration (at least several months old), brief feeding, and regular water changes, then the germ count should remain very low. The novice, who may not yet have the "feel" for good water, would do well to fall back on a UV lamp during the first weeks. This lamp should be suitably rated for the size of aquarium, in order to produce the desired effect. Another option is to use a bacterial starter in newly set-up tanks; In this case, of course, no UV lamp should be used as this would also destroy the useful bacteria. These bacteria will ensure that the metabolic by-products are broken down more rapidly. They will start working within a few hours. At present there are products of this type from several manufacturers on the market. It is easy to carry out a test to see if the purchased starter works. In an overcrowded aquarium with heavy feeding and no extra bacteria in a bio-filter, then sooner or later the water will be come cloudy. The fishes with show their discomfort by clamping their fins and becoming timid. A good bacterial starter, at the correct dose, will produce clear water within a few hours and the fish will no longer exhibit any symptoms of the problem. Regular re-dosing is necessary in tanks without a bio-filter, and this will work well as long as we keep up the treatment. Such aquaria are very unstable, but sometimes very useful where help is limited. Despite having bacteria at work in my bio-filters, after each water change in heavily populated aquaria I use one of these far from cheap products as many important bacteria are removed during the water change. If no food is offered for a day then the system will have a chance to return to satisfactory function. In heavily populated tanks the the mechanical filter should be cleaned out several times a week. Such filters should be easy to clean, as otherwise the job is too often postponed.

Temperature

Another avoidable mistake, which may cause losses in the long term, is the habit of keeping the temperature at 24°C through ignorance or for reasons of expense, although to some extent we can fit the fishes in with our budget in the short term, during the winter, with a simultaneous reduction in feeding. This brief seasonal rest permits the fishes, sadly often purchased only to make money, a chance to recover. Any seasonal change of temperature should take place over several weeks, as in nature such changes are likewise very gradual. As already mentioned, this measure should be temporary and not overdone. In exactly the same way, we should not keep these cichlids constantly at high temperatures (28° - 30°C), as this may seriously shorten their life expectancy. I have recorded a water temperature of 28 - 30°C at the beginning of March in Ndole Bay in Zambia. A seasonal change in temperature is natural for the cichlids from the upper zones of the rocky littoral in Lake Tanganyika, and should thus also be implemented in the aquarium.

Decor during acclimatisation

Don't use any decor during the acclimatisation phase, unless your aquarium is too large. The strongest fish will establish a territory too quickly and tyrannise the other occupants, and this can often lead to losses.

Diseases in the aquarium

Apart from a few exceptions, our fishes become ill only if we give them reason to do so. If the conditions each genus requires are provided – and that is often neither particularly expensive nor difficult – then problems rarely occur. Many fishes – and this has been known for a long time - constantly carry diseases around with them, but succumb only when conditions drop below the basic minimum for the individual, weakening its immune system; and the parasites are there waiting for what are, for them, better conditions. Sometimes I get the impression that some aquarists spend more time – unknowingly, of course – creating and maintaining ideal conditions for pathogens than is good for their fishes. Then, when their fish fall ill, many are reduced to panic and generally make the wrong decisions. In the course of time they become regular customers at the local pharmacy. Very often they acquire numerous medications that should normally be prescribed for the sick fish by a veterinarian, and only after an accurate diagnosis. Because they generally wait too long, not spotting the symptoms right at the start, the outlook for the fish is often very bleak. On the basis, "double the amount, double the effect", they do not follow the dosage on the packet but use double or even several times the stated dose. And then it is a case of, "The operation was successful, but the patient died."! But some operate on the reverse basis, constantly adding a low dose of medication to the tank – and this is far more dangerous for the aquarium hobby in general. There are already resistant strains of some pathogens, and this makes it difficult or even impossible to combat these diseases using traditional methods. All too often medications are used where none are needed. It is not uncommon for fishes to become ill only after unnecessary treatment or overdosing. I wouldn't even like to guess how many fishes die from kidney and/or liver failure months after incorrect treatment. It is essential to have the sick fish diagnosed by a specialist in fish diseases, and if this is impossible because of the distance involved, then – after making the necessary arrangements with the "fish doctor" by phone - it can be sent by overnight courier. The aquarist will then be sent the results the next day, by telephone in urgent cases, along with a suggested plan for treatment. This is the easiest method of getting rid of a disease. The cost will be well worth it, compared to the value of a tank of *Tropheus*. There are a number of good books on fish diseases, and every aquarist should have such a book on his shelves.

A professional diagnosis is indispensable in saving the fish population and avoiding wasting money on a variety of medications that fail to help. Unfortunately, it frequently happens that the aquarist waits too long, by which time the fishes are often so badly affected that the eventual treatment is no help. It also unfortunately often happens that the primary pathogens are no longer present or are not (cannot be) diagnosed. Often such weakened fishes succumb to tuberculosis, which is then diagnosed. Tuberculosis can normally gain a foothold only in weakened fishes. The actual disease remains unrecognised and thus untreated. This often happens in *Tropheus*.

It is not unusual for them to be infested with protozoa such as *Trypanosoma* (blood parasites), *Cryptobia* (very dangerous), or *Hexamita*. In bad cases more than one pathogen may be involved. Once we have saved the fishes, then we must find out how the disease was able to strike. There are really only two possibilities. If the fishes fall ill after several months, then we must be self-critical and seek the help of an experienced aquarist to find out the cause(s) of the problem. Often this will be merely mistakes in maintenance and feeding, which are relatively easy to rectify. If, however, the fishes become ill during the first days after arrival, then this means we didn't study them closely enough at the dealer's or importer's. In order to avoid such unfortunate events, I always collect my fishes as soon as the importation arrives, or at most a few days later. Ailing fishes will usually have been weeded out on collection or at the latest before despatch by the exporter. I have rarely had problems with freshly imported fishes. All new acquisitions, whether wild imports or tank-bred, without exception should be placed in a quarantine tank. This tank should not be tucked away in some dark corner, but be at viewing level and have normal lighting and filtration. A small internal sponge filter is perfectly adequate. As already mentioned, a good "algae lawn" in the aquarium will be beneficial. It is advisable to dispense with decor objects, or to have at most a large rock (without caves) in the middle of the tank. This will prevent the formation of territories and keep aggression within bounds. Food should be withheld for the first day or two, as in fishes stress also affects the digestive tract. On the following days they can be fed with small amounts of easily digested foods such as Daphnia, Cyclops, and freshly hatched Artemia, plus small portions of flake and granules. Daphnia are often intermediate hosts for various fish parasites, and using this food live creates a high risk of infecting our pets. We should never run this risk, as to do so endangers our entire fish population. Even the smallest garden pond is visited by tame wild ducks and can be infected by their droppings. Another major danger, which even today is rarely taken into account, is infection via frozen food. Cyclops and Daphnia are known to sometimes contain the intermediate stages of various fish parasites, which complete their life cycles when consumed. It has only recently been discovered that some developmental stages of these parasites can be guaranteed dead only at temperatures of -70°C or lower. Their chance of survival should be lower in properly quick-frozen food than in cheap products from home and abroad. But there is still a certain residual risk. Every aquarist and dealer must, of course, decide for himself how valuable his fishes are and the degree of risk he is prepared to take. But you have been warned!

Regular deaths, spread over several months and with no identifiable cause, or diagnosed diseases that recur repeatedly despite appropriate treatment, are reason for suspicion. During my activities that led to this book, I repeatedly experienced losses, despite quarantine, that affected all species. Even odd individuals of relatively hardy fishes such as the Lamprologini group were affected. Post-mortem examination has revealed nothing tangible to date apart from the usual diseases such as various parasitic protozoa of the skin, gill flukes, and other relatively harmless maladies. After Cyclops and Daphnia were removed from the menu the incidence of disease decreased over a period of many weeks and eventually ceased completely. For this reason Cyclops and Daphnia fell under suspicion, as I used these foods only sporadically in the absence of regular availability from my supplier. The diseases were similarly sporadic. Since I stopped feeding these crustaceans my only losses have been occasional suicides that have escaped the tank through the tiniest of gaps or my carelessness, and eventually drawn attention to themselves a few days later by their unmistakeable aroma.

Another, perhaps even more important point is the apparent extreme sensitivity of Aufwuchs-feeders in particular to germs. In individual cases it can certainly be very helpful to know whether these were generated in the tank or came in with the mains water. Overpopulation of the tank, coupled with insufficient biological filtration and inadequate water changes, can cause this problem, but unsuitable mains water has driven many an aquarist to despair, and ultimately a change of hobby.

It will be beneficial if the quarantine tank is also to be the future home of the new fishes, as this will avoid the stress of another move. The decor can be completed after the acclimatisation period.

After a few months the general condition of the fishes should have stabilised completely. Only then will the fishes show their previously hidden splendour, and their owner will be astounded at the brilliance of the colours of many populations. Once the fishes have attained this state, then it will be rare for them to fall ill, and the aquarist will – if he is able and willing – have the pleasure of enjoying their special behaviour and their fantastic colours for 10 years or more. I personally could never become bored with a group of *Tropheus* sp. "Kirschfleck", *Petrochromis* sp. "Kaiser", or the little Tanganyika goby cichlids, and every day I am captivated anew.

Prophylactic treatment

Ask the importer or previous owner whether any medication has been carried out. Don't treat the newly-purchased fishes with the usual strong aquarium-scene medications on a "just-in-case" basis. That will do more harm than good. If you find yourself confronted with any insoluble problem, turn to an experienced aquarist whose tanks are well-maintained and whose fish are healthy. That is all that matters. He doesn't necessarily have to be an old hand. Listen to his advice and let him explain where you have gone wrong. If in doubt, a fish can be sacrificed and sent to a "fish doctor".

When you have purchased new fishes, the water in the transportation bags should be slowly mixed with your water. The pH may often be very different. Once the fishes are acclimated to the water, they should be given a salt bath in the bags, using a dosage of 0 - 15 g salt per litre of water. This short-term bah should last no longer than 20 minutes. A longer salt treatment later on is best avoided. A short-term salt bath remains the best and simplest method of ridding newly-purchased fishes of harmful ectoparasites, and guarantees that no new external parasites are introduced into the existing fish population. If the fishes turn on their sides then the treatment must be terminated immediately, but in general the cichlids of the Tropheini group will tolerate this treatment without problems. The transport water must be thrown away after treatment.

Of course it is better if the newly-purchased fishes can be placed in a quarantine tank for the first weeks, and any further treatment necessary can be carried out there. It is best not to "pollute" the established tank with medications. Do not offer the fishes any food – apart from the algae in the aquarium – during the first few days, as they must first recover to some extent from the stress of transport.

If the fishes are correctly and expertly collected, held over, shipped, and acclimatised, then they will live longer lives in our aquaria than in their natural environment. During the selling-in period it will become evident whether the importer has done his job properly. Of course, if our existing fish population is infected (a situation best avoided) then no blame attaches to the dealer. For this reason aquarists and importers should treat their tanks prophylactically against the most commonly encountered diseases. If an importer's tanks contain a pathogen that takes weeks or months to break out into visible disease, or may even be chronic in its effects, then this can cause huge long-term damage, plus often the disease may not be recognised as such. Various flagellates are prone to infest Aufwuchs-feeders such as *Tropheus* und *Petrochromis* species, and many an aquarist has fallen foul of this problem. The importer and dealer must disinfect all tanks as they fall empty in order to avoid new arrivals being infected in turn. Even a wet hand can transfer many pathogens from tank to tank.

Because recently increasing numbers of *Tropheus* have been

succumbing after weeks or months to an as yet unidentified disease, I have evolved the following "just in case" treatment procedure, which has worked very well:

a) A short-term bath in the transportation bags using a medication containing nifurpirinol following the stated dose.

b) A salt treatment (3 g/10 l at hardness values of more than 12°dGH in the quarantine tank, with a concluding water change.

c) Treatment a few days later against *Hexamita* and other similar endoparasites using a suitable medication at the stated dose.

Recommended reading on fish diseases will be found in the appendix at the end of this book.

Floating medication bags have proved particularly useful for medication. followed by a complete change of water. Water from an established tank can be used (the tank gets a partial water change at the same time), provided, of course, that the resident population is free of disease, as otherwise the entire process is a waste of time. It is strongly advised that the *Hexamita* treatment is repeated after two weeks. The same applies with *Ichthyophthirius*. Prophylactic treatment against white spot (Ich) should be administered twice a year.

An effective method of treatment is to get the medication into the fishes. A solution can be made and sprayed over food sticks or flake and allowed to soak in. Allow the food to dry out to some degree, and then feed it to the fishes. It is best to treat both the fishes and the water. When the medication is administered via the water, it is always very important to remove it completely from the water after treatment. This can be achieved via a complete water change or by using filter carbon. I have always performed a total water change instead of using carbon, to be sure, as I have never been able to check that the carbon does the job properly. The organs of the fish take time to rid themselves of the medication. Many a fish has died of kidney or liver failure rather than disease, following overdosing with medications Treatment must always follow the specific instructions for the medication. It is advisable always to have a white spot treatment handy as this disease typically appears when the shops are shut! There are treatments that do not colour the water and are harmless to the bacterial flora of the tank.

Cameron Bay. photo: P. Schupke

The genus *Tropheus*

At the beginning of the 1990s my hobby brought me back to the genus *Tropheus* again. One of the reasons was the numerous new colour forms that were available, but also old friends such as the "Kirschfleck" and the "Kaiser" impressed me with their almost unsurpassable colour and health. At that time *Tropheus* were being imported from all four of the countries bordering Lake Tanganyika, in large part due to a serious commitment on the part of a number of importers. The success in what was then Zaire is particularly worthy of mention. A lot of personal effort and money went into that eventual success. Anyone familiar with Zaire will know that there is always a high level of risk involved there. The efforts of the local collectors are no less worthy of mention. Pierre BRICHARD had already made numerous investigatory trips to the west coast back in the 1980s, publishing his experiences in his book, "Tanganyika Cichlids". Additional teams of varied nationality followed in the footsteps of Brichard, and it is them we mainly have to thank for some of the less spectacular forms also finding their way into our aquaria. On the basis of the numerous forms that gradually supplanted my much-loved featherfins in my aquaria, together with details of their provenance, I was able to put together an increasingly complete picture of the distribution of various lineages (subspecies, species) and how they bordered or overlapped one another. Common features seen in various forms led to an even more exact picture. These common features (including base colour, secondary colour, markings, fin proportions, etc), as well as the distribution patterns, indicate that the genus *Tropheus* contains more species than the six currently described taxa.

After many years of intensive research and study of almost 90 different populations, I have come to the conclusion that the fishes currently assigned to the genus *Tropheus*, both described and undescribed, should be split into at least 10 different lineages. There may be more. Whether these lineages represent subspecies, species, or perhaps even new genera, requires further research, including genetic and other studies. But without doubt it is the living fishes, whether in captivity or in the lake, that are the most important source of information.

I regard the observations published here as a sort of inventory, and they are certainly not the final word on the subject. They are also meant to provide an overview of the current state of our knowledge.

Before we go into the individual lineages and populations, an explanation of my system of categorisation is necessary.

Starting in the north, the black *Tropheus*, which have many features in common with the described taxon *Tropheus moorii*, are assigned to lineage 1. The number to the right of the dot represents one of the populations in this lineage. For purposes of example we will take Tropheus sp. "Kiriza". This population bears the number 1.8. The first number denotes the lineage (in this case lineage 1), and the second is the number of the population (in this case number 8). In cases where there is a third number after the population number, this denotes a sub-population within that population, eg *Tropheus* sp. "Lueba" is 1.51. The *Tropheus* in the northern part oft the distribution of the lineage exhibit a yellow stripe on a black background or are completely black. In the southern distribution area all individuals are black

Lineage 1: similar to *Tropheus moorii* BOULENGER, 1898
Base colour black; secondary stripes red, orange, yellow; adults do not exhibit transverse bands; fins always base-coloured, only the dorsal sometimes with secondary colour; posterior edge of caudal slightly concave; 5 anal spines. Sometimes found syntopic with lineages 2 and 5.

Lineage 2: comprises *Tropheus brichardi* NELISSEN & THYS VAN DEN AUDENAERDE, 1975
A heterogenous lineage, highly reminiscent of lineage 1. The main reason for lineage 2 being regarded as separate, as here, is the research of STURMBAUER and MEYER (1992). They postulated that *Tropheus* from Bemba were closely related to *Tropheus brichardi* from Nyanza Lac. Base colour dark brown rather than black as in lineage 1; fin colour as base colour; an orange band; indistinct or faded brown-orange stripes; a whitish saddle-spot or no markings on the dorsum; beige stripes only in juveniles; 6 anal spines. Sometimes found syntopic with lineages 1, 3, and 5.

Lineage 3: probably undescribed species
The characters of lineage 3 make it easy to differentiate from the other lineages. Base colour ranging from dark grey to dark brown; yellow secondary colour apparent in all fins except the caudal and as stripes on the body; only dominant individuals temporarily lose their stripes; 6 anal spines. Sometimes found syntopic with lineages 2,4,5,110 and 11.

Lineage 4: comprises *Tropheus polli* AXELROD, 1977 and *T. annectens* BOULENGER, 1911
Lineage 4 is very easy to differentiate from other *Tropheus* lineages. All populations have greatly prolonged caudal points, like pennants. Base colour dark grey to dark brown; subordinate individuals with clearly visible stripes; dorsal yellowish-grey; caudal fin deeply indented, its tips prolonged as filaments when full-grown; large ultimate size; juveniles with highly contrasting stripes and a more or less blue eye-ring, and very similar to those of lineage 9; 4 anal spines. Sometimes found syntopic with lineages 3, 5, 8 (?), 10 and 11.

Lineage 5: probably undescribed species
Lineage 5 is distinguished by its greenish base colour; yellowish to yellowish-green secondary colour; blue iris; rather longer and more strongly indented caudal fin with two short points; juveniles with yellowish stripes, yellow colour in anal fin and on throat and belly, spot or band posterior to operculum; 6 anal spines. Sometimes found syntopic with lineages 1, 2, 3, 4, and 10.

Lineage 6: probably undescribed species
Lineage 6 exhibits a brown-beige to brown-reddish base colour; underside light; distinctive markings in the form of stripes on the snout and forehead; bluish to bluish-grey colour in the dorsal, anal, and ventral fins; a small orange area posterior to the operculum and pectorals; 6 anal spines. This lineage is found adjacent to lineages 4,7, 8 and 12.

Lineage 7: probably undescribed species
Lineage 7 possesses a brown-red, red-brown to ochre-brown

base colour; dorsal, anal, and ventral fins with a secondary coloration of dark red, light red, orange-red, and yellow; in all cases stripes on forehead, not as prominent as in lineage 6; juveniles with narrow, light-coloured, fragmented stripes, often also still clearly visible in adults, especially on the back; caudal slightly indented; 6 anal spines. This lineage is found adjacent to lineages 6 and 8.

Lineage 8: comprises *Tropheus moorii* BOULENGER, 1898, and *T. kasabae* NELISSEN, 1977
Lineage 8 is very variable because of its wide distribution. Common features of all populations are fine, narrow, fragmented transverse bars, often only a single scale wide, sometimes with even narrower fragmented stripes in the intervals; a more or less lighter lower half to the body; posterior caudal edge almost straight-cut; residual traces of stripes always present on the dorsum; many populations exhibit fine dots on the head; females and non-territorial males with stripes; juveniles with even narrower stripes in the intervals; 6 anal spines. This lineage is syntopic with lineages 4, 6, and 9.

Lineage 9: probably undescribed species
Lineage 9 is characterised by a dark grey-green to dark brown-red base colour; a striking pattern of broad stripes; iris blue; dominant individuals lose their stripes, depending on locality either completely or with 2-3 yellow stripes remaining posterior to the gills and used as a signal in courtship and territorial display; fins sharing body base colour; all juveniles with an extraordinarily contrasting pattern of whitish yellow to yellow or orange stripes, resembling juvenile *Tropheus polli* except for the forked tail; caudal edge only slightly concave; usually 6 (5) anal spines. This lineage occurs syntopic with lineage 8 along a section of coast several kilometres long.

Lineage 10: comprises*Tropheus duboisi* MARLIER, 1959
Base colour black; head bluish; a more or less yellowish, yellowish-white, or white band of variable breadth; caudal slightly concave; juveniles with a very striking coloration, black with numerous light spots; usually 5 anal spines. This lineage is syntopic with lineages 2, 3, 4, 5 and 11.

Lineage 11: probably undescribed species
These cichlids – there is only one population – are known in the aquarium hobby as *Tropheus* "Ikola". In my view they do not fit into any other known lineage. They exhibit a brown-black base colour; a very broad lemon-yellow band on the body; a red marginal band on the dorsal; a dark iris with a red upper edge. It is at present unknown if this lineage is adjacent to lineage 9 at its southernmost extreme; in the northern part of its range it is restricted to deeper water by *T. polli* (STURMBAUER, pers. comm.).

Lineage 12: probably undescribed species
These *Tropheus* comprise a single form that again cannot be ascribed to any other known lineage. This population is known as *Tropheus* "Kirschfleck" or "double spot". It inhabits a stretch of coast about 70 km long and is adjacent to three other lineages (3, 4, 6 and 10) at its northern extreme.

Geographical distribution of the lineages

A very important argument in favour of the evolutionary discreteness of these lineages is the existence of contact zones between several lineages. Lineage 2 lives sympatric with *Tropheus* of lineage 5 in the extreme northwest in the Congo. The situation on the spot is dictated by the slow silting of the rocky biotope with sediment carried into the lake by the Ruzizi: lineage 2 considerably outnumbers lineage 5 in terms of individuals. The lower half of the Ubwari peninsula also harbours two lineages, specifically 1 and 2. The distribution of lineage 1 ends on the western shore at the southern tip of the Ubwari, which represents the southernmost point in its range. On the eastern shore the distribution of lineage 1 starts a few kilometres north of Rutunga and ends at Minago, where the lineage reaches the southernmost limit of its east-coast range.
The distribution of lineage 5 extends past the southern tip of Ubwari as far as Toa Bay, a few kilometres northeast of Kalemie. Approximately in the centre of the stretch of coastline between south Ubwari and Toa Bay, immediately opposite the outflow of the Malagarasi, lineage 5 is found alone. Thereafter it occurs together with lineage 3, which extends as far as M'toto.
At Rumonge in southern Burundi there are again two lineages living together, namely 5 and 1. BRICHARD speaks of the great variability of these *Tropheus*! The same situation obtains at Nyanza Lac, where we find *Tropheus* of the lineages 5, 3, and 2. The distribution of lineage 2 ends here. A few kilometres south of the Tanzania-Burundi border, at Gombe, lineages 5, 3, and *Tropheus duboisi* (lineage 10) live together. Lineage 3 continues as far as the mouth of the Malagarasi. Its distribution resumes some distance further south at Halembe, and ends again to the south Bulu Island, where it occurs with three other lineages.
Lineage 5 continues across the Malagarasi and currently (end 2001) is considered to end at the village of Halembe. There is a high probability that there are further populations to the south of this village. The site mentioned in the literature at Isonga, a village to the south of the Mahale Mountains, has proven to be false.
No less than 6 *Tropheus* lineages are found together at Karilani – numbers 10, 12, 4, 3, and newly discovered 2.
At the southwest lower extremity of the Mahale Mountains lies Cape Kibwesa, where Tropheus of lineages 12, 6, and 4 occur together. Recently Helmut LÖFFLAD by chance discovered a further population between Cape Kibwesa and Isonga; these fishes are very similar to *Tropheus* sp. "Kibwesa" and are assigned to the same lineage.
The *Tropheus* of lineage 6 belong to the *Tropheus* that are otherwise found only a long way away on the west coast at Zongwe and Kapampa. Opposite the Kungwe Mountains lineage 3 extends to M'toto in the Congo, and encounters lineage 4 in this region. The distribution of lineage 4 extends from Toa Bay to past Moba in the Congo, and on the eastern shore opposite from Cape Kungwe to well into the region north of Ikola. *Tropheus* of lineage 8 live at Pala, which lies almost central on the west coast of Lake Tanganyika, and their distribution extends a few kilometres south to Kapampa Bay, where they are replaced by lineage 6. Only in the far south, in the Zambian part of the lake, does lineage 8 re-appear, in south-eastern Nkamba Bay. Its distribution continues around the southern part of the lake to Kisambali, which lies in Tanzania opposite Kapampa Bay. At Kisambali lineage 8 encounters lineage 9, and the two are found together for a few kilometres.
The distribution of lineage 7 begins at the southern extremity of Kapampa Bay in the Congo, where it contacts lineage 6. The other limit of the range of lineage 7 is at the Nangu peninsula, where it encounters lineage 8, whose distribution resumes here. The Tropheus, that occur at Ikola cannot be assigned to any other known lineage and hence are placed in their own, lineage 12. The distribution of this lineage extends north of the village of Ikola as far as the angle in the coastline at Isonga. To the south, lineage 9 approaches to within a few kilometres of the site at Ikola.

Underwater reef in Nkamba Bay. photo: P. Schupke

The evolution of the genus *Tropheus*

Christian Sturmbauer and his colleagues have for some time been researching the phylogenetic relationships of populations of the genus *Tropheus* with the aid of DNA sequencing. They have been kind enough to provide us with a manuscript concerning their work for inclusion in this overview of the genus Tropheus.

The evolution of the genus *Tropheus*

Christian Sturmbauer*

The systematic classification of the genus Tropheus

The genus *Tropheus* currently comprises six described species: *T. annectens*, *T. brichardi*, *T. duboisi*, *T. kasabae*, *T. moorii*, and *T. polli*. As with many other cichlids of Lake Tanganyika, five of the six *Tropheus* species include a series of differently coloured populations, known to aquarists as geographical races or morphs. By far the majority of these populations are not formally described scientifically. During the last decade the genus *Tropheus* has several times been the subject of genetic investigation using molecular methods, providing interesting insights into its history and also a number of discoveries that run contrary to the currently accepted taxonomy. It is clear that the currently accepted taxonomic classification of the genus *Tropheus* does not represent the actual phylogenetic relationships involved. There are several reasons for this. Firstly, George Boulenger, who described the first *Tropheus* species in 1898, was unaware that hundreds of geographically isolated populations of these fishes existed, in some cases exhibiting considerable variations in morphology and coloration. Subsequently, as increasing numbers of populations were discovered, individual populations, found on closer examination to be clearly distinct from previously described taxa, were elevated to specific status. Thus at the present time we are confronted with the situation that the actual systematic diversity of the group remains to be established, as to date no representative studies in comparative anatomy have been performed in order to group the multitude of geographical races into closely related evolutionary lineages. So far attempts to define such groups have been based solely on similarities of coloration and markings. The classification used in this book is without doubt the most accurate devised to date on the basis of coloration and distribution. The most complete genetic study using molecular methods was published a few weeks before this book. Each method has its advantages and drawbacks: similarities in coloration may be coincidental, so that a classification of the genus using this criterion may be misleading. And even genetic characters have their limits. For this reason a reliable theory can be derived only from the synthesis of the results of numerous analyses of characters. In the text below the techniques of molecular phylogenetic research are explained, followed by an overview of the current state of our knowledge.

Reconstruction of evolutionary history using DNA sequencing

The genome of every life form contains not just comprehensive information on the development of the body structure, the organs, and all behavioural characteristics but also valuable data regarding evolutionary relationships. This is the case because over the passage of time increasing numbers of mutations accumulate in the genomes of individual species. Hence the degree of evolutionary relationship between two species can be calculated from the degree of genetic variation between them. By creating molecular phylogenetic trees, the genesis, that is the evolutionary history, of life forms can be reconstructed with hitherto unattainable accuracy. This reconstruction of the genesis of species is possible because the genetic information of all organisms since their origin is passed down from generation to generation. Every group which at a given point in time is able to share its genetic information via free partner choice, has a common evolutionary future. Such groups are termed populations. If a population voluntarily or involuntarily splits into two daughter populations, then thereafter the latter have their own separate evolutionary futures as they no longer exchange genes after becoming separate. The arbitrary occurrence of mutations leads to these daughter populations becoming increasingly genetically and morphologically distinct until eventually they pass an invisible boundary and become two separate species. If two individuals can no longer breed together successfully then an inter-species boundary now exists. In the final analysis, the genesis of new species requires a branching in the lines of inheritance by which genetic information is transferred, and thus the phylogenetic tree of all life is formed.

The genesis of new species via geographical separation

The majority of models for speciation require at least temporary spatial isolation, by virtue of which two or more populations can diverge genetically and morphologically until they split into reproductive communities that remain discrete even if they once again come into contact. The process of speciation is influenced by the interaction of certain biological characteristics of the mother species with environmental factors. Limited potential for expansion in a species is a biological prerequisite, so that a developing spatial separation between two habitats can become an insuperable barrier. It is easy to imagine that to *Tropheus*, which, as *Aufwuchs*-feeding cichlids, are so specifically adapted to life in the shelter of underwater rocks, a 2 km stretch of sandy coast can be an insuperable barrier. There are thus certain biological characters of the species which determine whether or not a given physical distance can act as a barrier to gene flow. It makes sense to assume that smaller distances should suffice in the case of specialists, while generalists can bridge wider gaps. The second determining factor in speciation is an alteration in the habitat. In this way barriers may arise and also disappear again. It is easy to see how a climate-induced lowering of the lake level in Lake Tanganyika could create a sandy coast which would in turn split a *Tropheus* population into two reproductive units that were separate from then on. The effects of fluctuations in lake level are often localised, but sometimes far-reaching, as, for example, after the splitting up of a single lake into several smaller ones or after almost complete desiccation.

*Christian Sturmbauer, Institut für Zoologie, Karl-Franzens-Universität Graz, Universitätsplatz 2, A-8010 Graz, Österreich,
E-mail: christian.sturmbauer@uni-graz.at

Sturmbauer, C. (2003), The evolution of the genus *Tropheus*. In: Schupke, P. AQUALOG Cichlids of Lake Tanganyika, Part 1: The species of the genus *Tropheus*. Rodgau, Germany. :26-29

The extent to which fluctuations in lake level can affect the future of a species is demonstrated by the genus *Tropheus*, for which to date some 70 populations or sibling species have been catalogued (about 120 in this book). *Tropheus* is one of the most popular aquarium fishes from Lake Tanganyika on account of its interesting behaviour and enormous variety of colour variants. Our genetic research using molecular methods indicates that this genus has remained virtually unchanged morphologically for about 1.5 million years, although large genetic and colour differences have developed between populations. A particularly interesting discovery is that *Tropheus duboisi* is not a *Tropheus* at all and is instead only distantly related to other members of the genus. This species is probably more closely related to *Simochromis* and *Lobochilotes* than to other members of its genus. In addition we have been able to demonstrate eight separate genetic lineages which in all probability represent distinct species (Figure 1). The pattern of genetic relationships between the geographical forms and sibling species leads to the conclusion that these specialised rock-dwellers expand predominantly passively, as *Tropheus* cannot bridge unsuitable stretches of coast by their own efforts. Fluctuations in the lake level are the "engine" that periodically causes rocky habitats to amalgamate and separate again. From the pattern of genetic relationships in the genus *Tropheus* at least two major waves of expansion can be deduced: the first, in which the newly evolved species, the "primeval *Tropheus*", very rapidly expanded into all the suitable rocky habitats of the lake, and then subsequently underwent genetic and colour changes in each continuous rocky zones. The distribution pattern of this "primary radiation" was subsequently overlain by a second wave of expansion, which was very probably triggered by a massive fluctuation in the lake level.

These "secondary radiations" then overlaid the existing distribution pattern more or less within localised limits. Populations that had formed during the second wave of expansion differ from those resulting from the primary radiation in that they remain appreciably closer genetically. A sudden drop in the lake level must thus have extinguished or amalgamated numerous populations in the shallow littoral zone of the lake, in order subsequently to trigger a new wave of expansion when it rose again. The level of Lake Tanganyika must have dropped so far at that time that it was divided into three smaller lakes corresponding to its three deep basins. Because during this spitting up of the lake continuous coastlines formed across the original lake, even species as strictly adapted to life in rocky biotopes as *Tropheus* were able to cross to the opposite side. Thus various genetically ancient lineages came into contact in the overlap zones of the lake basin, sometimes interbreeding, or sometimes instead occupying niches in the same habitat as new species. Our genetic analyses indicate the presence in some cases of two ancient genotypes in the same population, pointing to the mingling of two long-separated evolutionary lineages following secondary contact. The double-spot *Tropheus* along the Kungwe mountains (lineages E and F in figure 1) still carry the genetic traces of two ancient lineages: the first is relatively closely related to *T. polli* (lineage E), while the second lineage belongs to the group of *Tropheus* races from Cameron Bay and the adjacent Congo to round about Mvua (lineage F). The same applies to the population at Katoto, which contains 50% of the genotype of each of two main lineages that encountered one another at Katoto and interbred. This population forms the boundary between lineage A4 in the south-east of the lake and lineage G in the south-west (see figure 1). On other stretches of coast two or three *Tropheus* species live sympatric in the same habitat. These were no longer able to interbreed when they came together, as they had already passed the invisible boundary that defines a biological species

References

Sturmbauer, C., & A. Meyer (1992) Genetic divergence, speciation and morphological stasis in a lineage of African cichlid fishes. *Nature* Band 358, S. 578-581.

Sturmbauer, C., E. Verheyen, L. Rüber & A. Meyer (1997) Phylogeographic patterns in populations of cichlid fishes from rocky habitats in Lake Tanganyika. In: Molecular Systematics of Fishes (T. D. Kocher & C. A. Stepien, Eds.), Academic Press, Seite 97-111.

Sturmbauer, C. (1998) Explosive speciation in cichlid fishes of the African Great Lakes: a dynamic model of adaptive radiation. Journal of Fish. Biology Band 53, Supplement A, Seite 18-36.

Baric, S., W. Salzburger & C. Sturmbauer (2002) Phylogeography and evolution of the Tanganyikan cichlid genus *Tropheus* based upon mitochondrial DNA sequences. Journal of Molecular Evolution 56: 54-68

About the author

Christian Sturmbauer: Born in 1960 in Linz. From 1979 studied biology at the University of Innsbruck, Sponsion 1986, Promotion 1990, 1991-1993 Postdoctoral sabbatical at the Department of Ecology and Evolution at the State University of New York at Stony Brook, 1993-1995 Contract Assistent 1995-1998 Assistant Lecturer and head of the molecular genetics laboratory at the Institute of Zoology and Limnology of the University of Innsbruck, 1998 Habilitation, thereafter Extraordinary Professor at the University of Innsbruck, since 2002 Professor of Zoology at the Karl-Franzens University, Graz, Austria.

Sturmbauer, C. (2003), The evolution of the genus *Tropheus*. In: Schupke, P. AQUALOG Cichlids of Lake Tanganyika, Part 1: The species of the genus *Tropheus*. Rodgau, Germany. :26-29

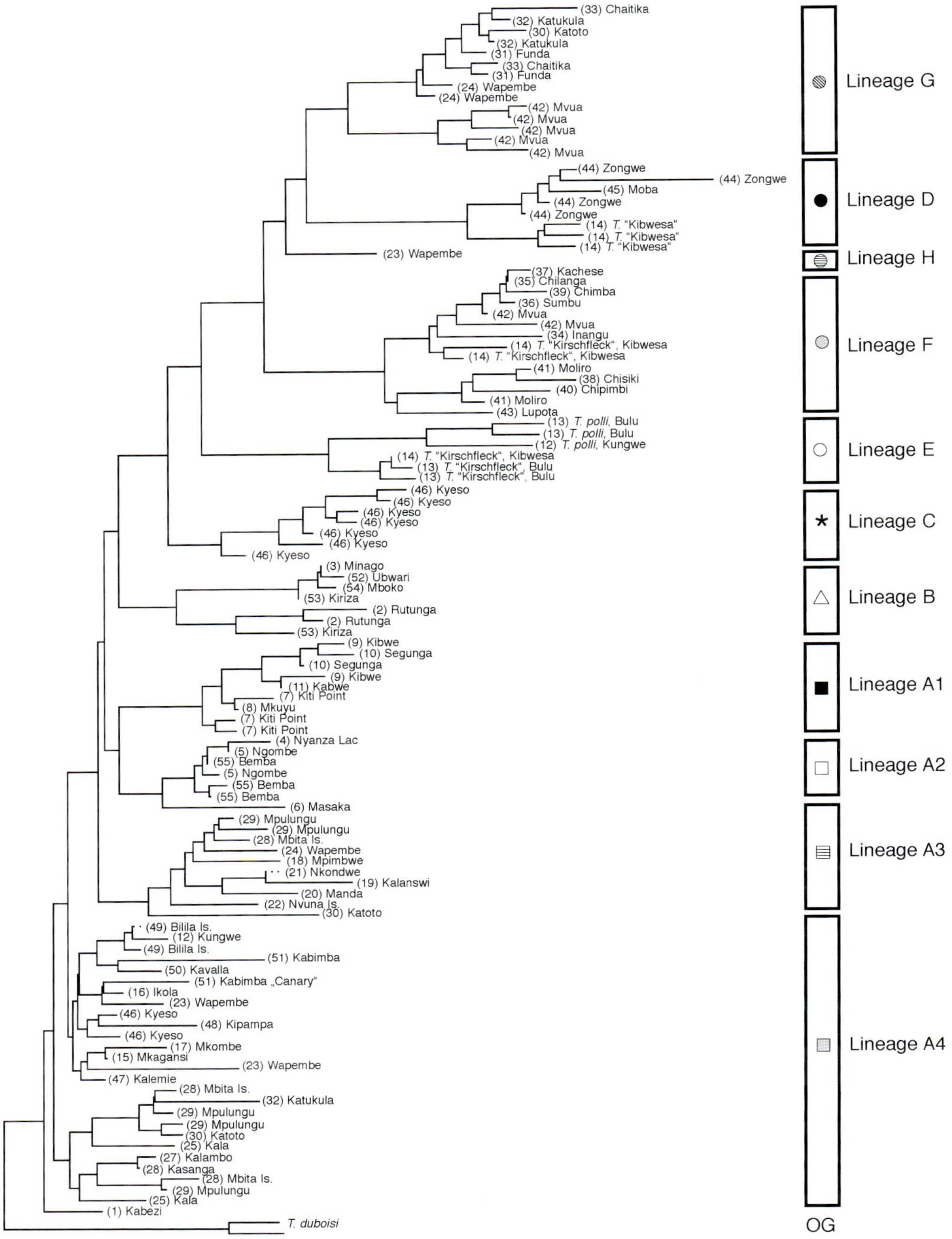

Phylogenetic tree and distribution map (opposite page) of 55 populations of the genus *Tropheus*. On the basis of this tree, eight genetic lineages (lineages A-H) are defined and denoted as blocks. The 55 populations studied are denoted on the map of the lake by the name of the settlement closest to the collecting locality. The current distribution of the eight genetic lineages is the result of at least two waves of expansion, with a high probability that these were triggered by major fluctuations in the lake level. In the course of the primary radiation the newly evolved species spread very rapidly to all the rocky zones, to subsequently evolve independently at each site. The distribution pattern was then overlain by at least two further waves of expansion (secondary radiations). Prior to the secondary radiation Lake Tanganyika must have dropped to some 600 m below its current level and been divided into three smaller lakes, such that two lineages were able to cross the central part of the lake. One of the eight genetic lineages (lineage A) was particularly successful during this second radiation and replaced numerous *Tropheus* populations resulting from the first.

Sturmbauer, C. (2003), The evolution of the genus *Tropheus*. In: Schupke, P. AQUALOG Cichlids of Lake Tanganyika, Part 1: The species of the genus *Tropheus*. Rodgau, Germany. :26-29

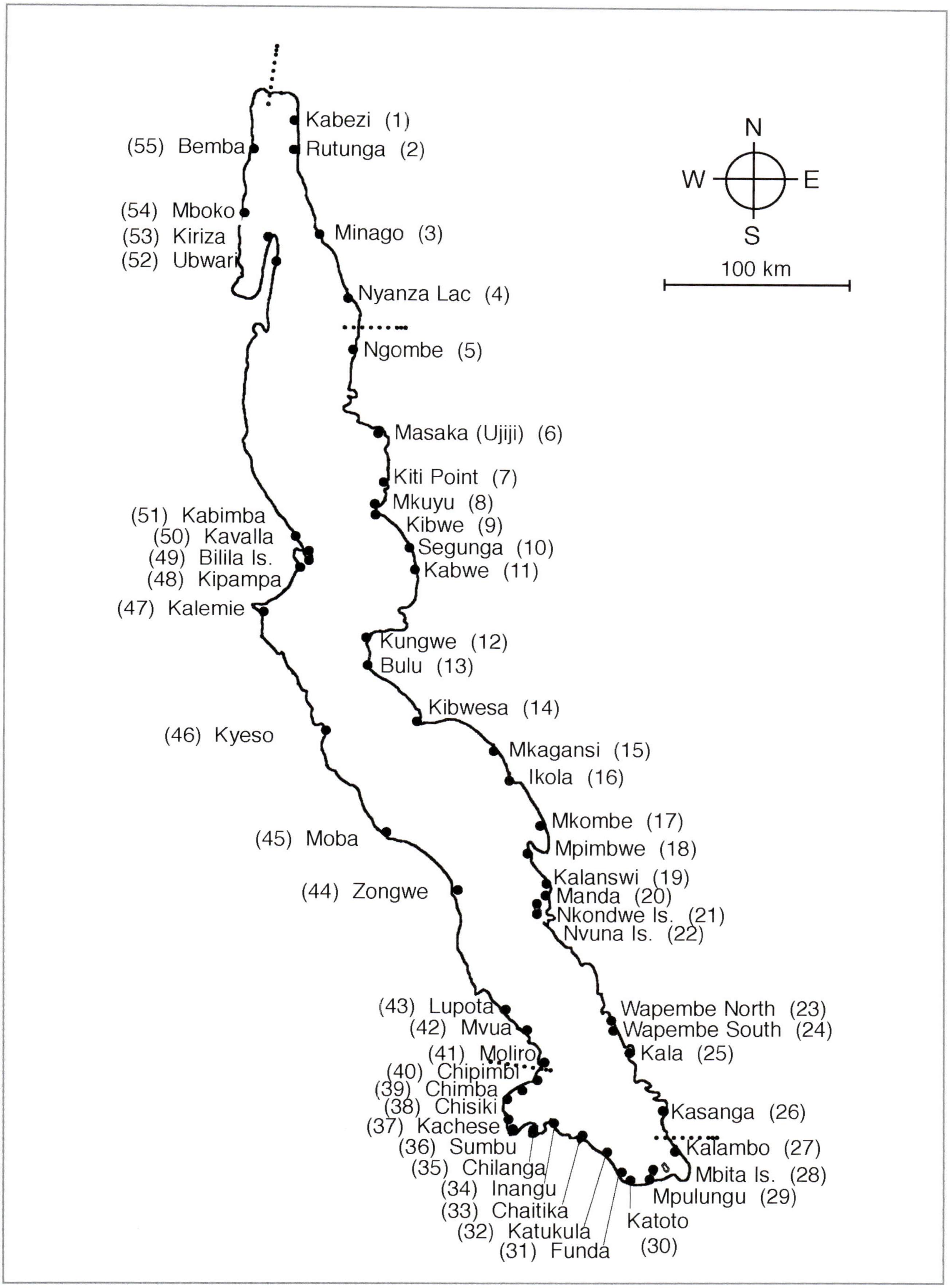

The taxonomic situation in the genus *Tropheus* BOULENGER, 1898

by Frank Schäfer

At this point it seems appropriate to discuss how the six described species of the genus *Tropheus* correlate with the numbered lineages used in this book.

The first species of the genus to be described was *Tropheus moorii* BOULENGER, 1898. It is simultaneously the type species of the genus *Tropheus* BOULENGER, 1898. The formal, valid, original description of both genus and species took place in BOULENGER, 1898a (see references, below). However, a more detailed description, with plates, took place in BOULENGER, 1898b. Because both publications appeared in the same year, there is frequently confusion regarding the correct publication date for the original description.

At the time of the original description BOULENGER knew nothing about the wealth of forms the genus *Tropheus* contains. He gives the type locality as Kinyamkolo. Today the whereabouts of this place are something of a mystery, as there is no longer any settlement or locality of that name. Because of the similar-sounding name of the missionary station Niamkolo at the centre of the southern end of the lake, NELISSEN (1977) assumes that the typical population of *T. moorii* should be sought in the region around what is today Mpulungu, which was subsequently founded near to Niamkolo church. HERRMANN (1997) furthermore points to a map in BOULENGER (1906), from which it is clear that the island nowadays named Kumbula Island was formerly known as Niamkolo Island and Mbity Rocks.

VON DRACHENFELS (1995) points out that the description of the stretch of coast provided by the collector, MOORE, is counter to the topography of the stretch of coast at today's Mpulungu: MOORE speaks of "craggy ledges", while the actual coast at Mbete Bay, where Mpulungu is located, is – at least now – edged with reed-beds. In addition, VON DRACHENFELS interprets MOORE's statement "Nearly all the new forms which I obtained were killed by dynamite from the craggy ledges of the west coast of the lake ...." as implying that the type locality of *T. moorii* should be sought along the south-western shore, and not in the extreme south of Lake Tanganyika. He also refers to a map that accompanies BOULENGER (1901b), which shows a "Kinyamkolo" on the south-western shore of the lake, but also makes the point that other locations shown on this map (Bete and Abercor (Mbala)) are clearly incorrect, which significantly reduces the value of the map. In my view the broad statement "Nearly all...." permits no dispute regarding the concrete data given by MOORE regarding the collecting locality of *T. moorii* at Kinyamkolo. So where does this Kinyamkolo lie, and which of the numerous *Tropheus* forms that we know today is the true *T. moorii*? BOULENGER received his information from MOORE. The maps that BOULENGER had available and those he used in his publications were undoubtedly not prepared by BOULENGER himself. Thus there is always a danger of an undetected transcription error. The direct statements of the author are far more reliable. And in the paper, "Third contribution to the Ichthyology of Lake Tanganyika. - Report on the Collection of Fishes made by Mr. J. E. S. Moore in Lakes Tanganyika and Kivu during his Second Expedition, 1899-1900" BOULENGER (1901a) makes precisely such a statement: on page 138 he lists the collecting localities, among them "Kinyamkolo, extreme south end of Lake Tanganyika". The description of the coloration of the type specimen is, moreover, very characteristic. Five of the specimens collected by MOORE were examined by BOULENGER, one of which had been prepared as a skeleton. The type material still exists today, with the skeleton (uncatalogued) and three of the syntypes in London (BMNH 1898.9.9.53-55) and the remaining syntype in Paris (MNHN 1898-0700). There are no grounds for the concern expressed by VON DRACHENFELS that the syntypes in the British Museum might belong to two different populations because BOULENGER subsequently included additional individuals, collected by the Cunnington Expedition on the Zambian west coast (Sumbu), in the description of *T. moorii* in his "Catalogue of the fresh-water fishes of Africa in the British Museum". BOULENGER's (1898b) description of coloration reads as follows: "Dark brown; a large bluish-white blotch on each side; belly reddish brown; fins blackish." Anyone at all familiar with the evaluation of the coloration of fishes in alcohol will know that details of colour (red, yellow, blue) are of no value. But although the colour contrast fades in the course of time (sometimes within a few years), in relatively recently preserved specimens the dark and light areas invariably contrast clearly. So there can be no question but that BOULENGER examined specimens of lineage 8, as this is the only lineage from the collecting area in question (regardless of whether actually the south or south-west) with a large light spot on the flank. Comparison with the drawing of one of the syntypes in BOULENGER 1898b permits us to specify the form in question even more precisely as one of the populations herein classified as 8.8 - 8.12.

On the basis of the combination of the arguments presented here regarding the type locality and the colour description provided by BOULENGER, it should in future be assumed that the specimens used by BOULENGER in the description of *Tropheus moorii* originated from the extreme south, where there are fishes of population 8.9. whose coloration accords perfectly with that described by him. The Mbete population of *Tropheus moorii* is thus the only one to which the name *Tropheus moorii* can unreservedly be assigned.

The situation is not much easier regarding the second taxon: In 1900 BOULENGER described the second species of the genus as *Tropheus annectens*. The type locality is given as Albertville (nowadays Kalemie), but there are no *Tropheus* there with only 4 anal spines, a character known only in *T. annectens* and *T. polli*. On the other hand it is very simple to assign it to one of the lineages, as the combination of sexual dichromatism, 4 anal spines, and a deeply indented caudal with long pointed tips is very characteristic. In this book these fishes are assigned to lineage 4. KONINGS (1999) synonymises the species *Tropheus annectens* and *T. polli*, arguing that there are no morphological differences between the two forms. That may well be (although I know of no detailed studies on the subject). KONINGS lists only one form with 4 anal spines from the west coast (from which *T. annectens* was described): that which inhabits a 15-25 km stretch of coast at M´toto. Although some doubt remains whether this M´toto population is precisely that which BOULENGER examined, ie *T. annectens*. The political situation in the Democratic Republic of the Congo makes it unlikely that the question of the existence of further populations of *T. annectens* will be answered in the near future.

The question likewise remains unanswered whether the extremely similar species *T. polli*, described from the east coast (type locality: South of Bulu Isl., 6°01´S, 29° 45´E, Kigoma district, Tanzania; in this book population 4.1.) is truly identical with *T. annectens*. However, in the light of everything that we know about *Tropheus* to date, this is improbable, because, as has already been shown, *Tropheus*, when geographically isolated, are inclined rapidly to form new phenotypes at least. Provided the east coast population is not the result of human activity, and did not travel from the west coast to the east (or vice versa) under highly unusual circumstances in the recent past (ie during the last 500 years), then it is far more likely that the west and east coast populations are different species than that they are identical. Until such time as conclusive research has been performed, it is wiser to regard them as two separate species. The west coast *Tropheus* with 4 anal spines must accordingly be assigned to *Tropheus annectens* BOULENGER, 1900 (population 4.3. in this book) and that from the east coast to *Tropheus polli* AXELROD, 1971 (population 4.1.) until a formal study can be undertaken using additional characters other than just fin counts and other morphometrics, and based on extensive collections. The minimum requirement, systematically speaking, for specific status is that the taxon should be distinguishable from other species without reference to its origin. This is the case with *T. annectens* and *T. polli*, even though the differences are admittedly only slight. As far as is known to date, in adults of the western species (*T. annectens*) the iris ring in adults is always black or grey, while in the eastern form (*T. polli*) it is always bright blue.

The other taxa of the genus *Tropheus* are far less problematical:

Tropheus duboisi MARLIER, 1959 was originally described from the population at Bemba (population 10.1). The four further local forms of lineage 10 known to date are all quite close to this species.

Tropheus brichardi NELISSEN & THYS VAN DEN AUDENAERDE, 1975 originates from Nyanza Lac, 4°21´S, 29°36´E, in Burundi. The type population is classified as 2.1 in this book. This form is quite unmistakeable and its specific status has so far been questioned by no-one. However, several further populations in lineage 2 are generally very differently coloured, which makes it likely that additional species will be described within this group in the future.

Tropheus kasabae NELISSEN, 1977, from Kasaba Bay, 8°31´S, 30°42´E, Zambia, has been synonymised with *T. moorii* by various authors, eg KONINGS, 1999, inter alia because it is unclear which population NELISSEN regarded as "typical" *T. moorii*. However, because *T. kasabae* undoubtedly differs considerably from population 8.9., herein demonstrated to be the typical population of *T. moorii*, this "technical" criticism has no significance. In Kasaba Bay occur "Yellow Rainbows" of the lineage 7.

It is impossible to say at this time precisely how many species or subspecies of *Tropheus* actually exist. The six taxa described so far can be differentiated using the following key (which applies only to full-grown (adult) specimens):

1 – Anal normally with 5 -7 spines 3
- Anal normally with only 4 spines 2

2. – Iris ring black or grey .... *T. annectens*
- Iris ring blue .... *T. polli*

3. A large yellow spot, not in the form of a band, on the flank of adult specimens. The spot begins posterior to the pectoral insertion and ends before the posterior end of the dorsal fin. . . .. *T. moorii*
- Other coloration 4

4. Head blue, rest of body blue-black. A white or yellowish transverse band on the body, always beginning below the anterior dorsal, never extending into the dorsal, which is without markings and the colour of the body .... *T. duboisi*

- Other coloration 5

5. Distinct sexual dichromatism: males without, females with stripe pattern; males uniform dark with a saddle-like spot below the anterior third of the dorsal, not extending into the fin. .... *T. brichardi*
- No sexual dichromatism, males and females largely identically coloured. No saddle-spot on dorsum. . .*T. kasabae*

This taxonomic situation is, of course, highly unsatisfactory. Some 120 different *Tropheus* phenotypes are known to date, and no extensive comparative anatomical studies have been performed. Peter SCHUPKE herein distinguishes 13 lineages, but the number of species may well be rather larger – from the data I have available I would estimate some 14-16. The different colour populations of *Tropheus*, which are connected by intermediate forms, are probably subspecies and should be described as such when the long overdue revision of the genus takes place. We must hope that the political situation in Burundi and the Democratic Republic of the Congo stabilises before too long. Mainly for the sake of the people that live there, of course, but also so that the great *Tropheus* mystery, which has already occupied scientists for more than 100 years, can finally be solved.

References:

Axelrod, G. S. (1977): A new species of Tropheus (Pisces: Cichlidae) from Lake Tanganyika. J. L. B. Smith Inst. Ichthyol. Spec. Publ. No. 17: 1-12

Boulenger, G. A. (1898a): Report on the fishes recently obtained by Mr. J. E. S. Moore in Lake Tanganyika. Proc. Zool. Soc. Lond. 1898 (pt 3): 494-497

Boulenger, G. A. (1898b): Report on the collection of fishes made by Mr. J. E. S. Moore in Lake Tanganyika during his expedition, 1895-96. Trans. Zool. Soc. Lond. v. 15 (pt 1, no. 1): 1-30, Pls. 1-8

Boulenger, G. A. (1900): Matériaux pour la faune du Congo. Poissons nouveaux du Congo. Sixième Partie. Mormyres, Characins, Cyprins, Silures, Acanthoptérygiens, Dipneustes. Ann. Mus. Congo (Ser. Zool.) v. 1 (fasc. 5): 129-164, Pls. 48-56

Boulenger, G. A. (1901): Les poissons du Bassin du Congo. Publication de l´État Indépendant du Congo. Bussels. i-lxiii + 1-532, 25 pl., 1 map.

Boulenger, G. A. (1915): Catalogue of the fresh-water fishes of Africa in the British Museum. Vol. 3: i-xii + 1-526.

Eschmeyer, W.: Catalog of Fishes (On-Line, ver. November 12, 2002)

Herrmann, H.-J. (1997): Tropheus. Dähne Verlag, Ettlingen, 112 pp.

Konings, A. (1999): Tanganjika-Cichliden in ihrem natürlichen Lebensraum. Cichlid Press, El Paso, 272 pp.

Marlier, G. (1959): Observations sur la biologie littorale du Lac Tanganyika. Rev. Zool. Bot. Afr. v. 59 (nos. 1-2): 164-183, Pls. 1-2

Nelissen, M. H. J. and D. F. E. Thys van den Audenaerde (1975): Description of Tropheus brichardi sp. nov. from Lake Tanganyika (Pisces, Cichlidae). Rev. Zool. Afr. v. 89 (no. 4): 974-980.

Nelissen, M. H. J. (1977): Description of Tropheus moorii kasabae n. sspec. (Pisces, Cichlidae) from the south of Lake Tanganyika. Rev. Zool. Afr. v. 91 (no. 1): 237-242.

v. Drachenfels, E. (1995): Was ist Tropheus moorii Boulenger, 1898? DCG-Info 26 (11): 255-260.

Type specimens of *Tropheus moorii* (above, BMNH 1898.9.9.53-55) and *T. annectens* (below BMNH 1900.12.13.31). photos: F. Schäfer

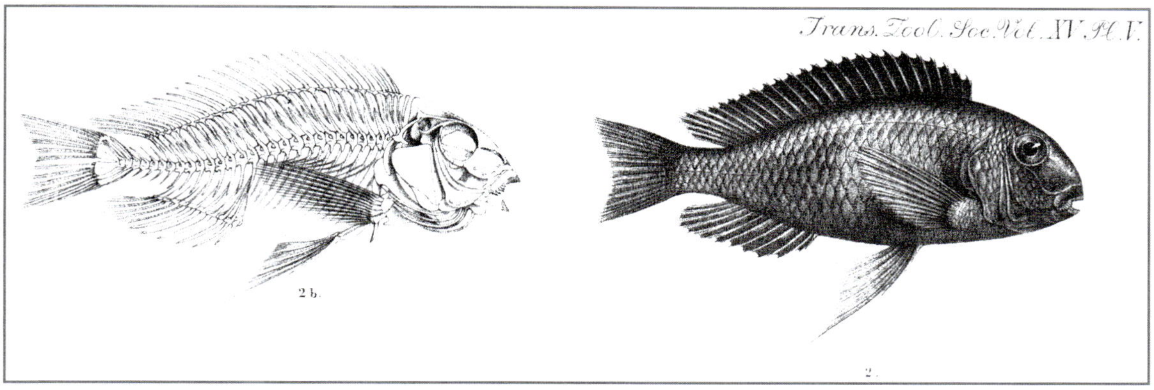

Left: Figures of two type specimens of *T. moorii* from BOULENGER, 1898b.

An attempt to interpret the distribution of the lineages

I was surprised how often my observations and researches correspond with the findings of Dr. Sturmbauer et al., sometimes in the smallest details. However, it should not come entirely as a surprise that very different methods and techniques also occasionally lead to conflicting results. Thus research into mitochondrial DNA allows us a glimpse into the past, revealing the immediate degree of relationship between any two specimens studied, and thereby permitting conclusions to be drawn regarding the phylogenetic relationships between different populations.
On the other hand, the study of living fishes and their habitats allows us a glimpse of the present. As the descriptions of the individual lineages show, living *Tropheus* exhibit characters such as colour and colour patterns that are no longer recognisable in dead specimens.
The morphometric differences between the various *Tropheus* are only slight, at least as far as is known today. Places in the lake where different populations are found together offer good opportunities for study. It is, however, not always easy to determine in the field whether *Tropheus* of different lineages are involved. On superficial inspection, the fishes of different lineages in a single locality are very similar. This has less to do with phylogenetic relationships than with convergent optimisation of camouflage to the habitat in question. But if the phenomenon of lineage distribution is studied in relation to the various different lake levels that may have occurred in the history of the Earth, then this provides an interesting starting point for future research.

In the course of its history, variations in water level have led to the lake being split into three smaller lakes. Perhaps a fourth such lake existed south of what is now Lake Kivu, which at that time had no outlet, but is today filled with the sediment of the Ruzizi. The process is thought to have occurred more than once in the history of the lake. These events brought a number of populations into contact to form new populations, or moved them further apart, leaving traces in the genetic material of these "migrants". The third main lineage mentioned earlier has left its traces all over the lake and founded new lineages (eg lineage 8). The recently discovered *Tropheus* sp. from Kabezi (population 13.0), which lies somewhat to the south of Bujumbura, represents a very ancient lineage that may perhaps have been more widespread in the hypothesised fourth, now silt-filled, lake. The sediment carried by the newly created Ruzizi would have forced the local *Tropheus* further and further south, and where the topography of the land made this impossible, they would have died out. Today something similar is happening to the populations living along the west coast in the north of the lake. If this hypothesis is correct, then the small population of the *Tropheus* sp. "Kabezi", threatened with extinction sooner or later, is a relict from earlier times. It has had no contact with other populations, as from a genetic viewpoint it is very homogenous. The same applies to the *Tropheus* of lineage 1. The genes of lineage 12 (*T.* "Kirschfleck") are very heterogenous particularly in the southern part of its range, indicating an influx of several other populations over the course of time. This led to a phylogenetic relationship between this lineage and lineage 7, and in particular to *T.* sp. "Lupota". In the northern part there is a genetic correlation with the lineages that occur south of Kigoma. This may have taken place when the northern and central lakes became joined.

What is now lineage 8 is also very heterogenous, as it originated from several sub-lineages of two main lineages in the south. Thus a component of lineage 8 derives from a splitting of lineage 3, to which belong *T.* sp. "Kavala" and *T.* sp. "Kalemie". Another component derives from another sub-lineage of the same main lineage (3) – the Wapembe and Mpulungu forms. The Chaitika form originates from another main lineage, introducing another component into lineage 8. The result was the current lineage 8 phenotype. As a result of the fluctuating water level, in the south too there is a genetic correlation between the *Tropheus* from Wapembe and those from Chaitika, and vice versa. The populations from the south are likewise assigned to this related group.

The Kibwesa *Tropheus* (lineage 6) probably stems originally from the west coast. A subsequent rise in lake level must then have fragmented the population at the southern Kungwe Mountains, where it came into contact with the local *T.* "Kirschfleck" (lineage 12) or its ancestors. Thus each received genes from the other. Today they live sympatric and represent two different species which are, so to speak, cousins.

The present *Tropheus* of lineage 5, along with *T. duboisi*, are the oldest *Tropheus* in the lake. They may have been distributed all over the northern lake at a very early stage, and have been only partially successful in warding off the influx of other populations. They do not hybridise with other *Tropheus*. This may perhaps be a function of their behavioural patterns no longer being comprehensible to the incoming *Tropheus*, which thus did not cross with them. Nowadays they have been restricted to the southern part of the northern basin by the *Tropheus* of lineage 3. From Kunanwa to Kasimia Bay they are still the only form, and exhibit a high population density. Across the lake, on the Tanzanian coast between Cape Kabogo and the Malagarasi, where

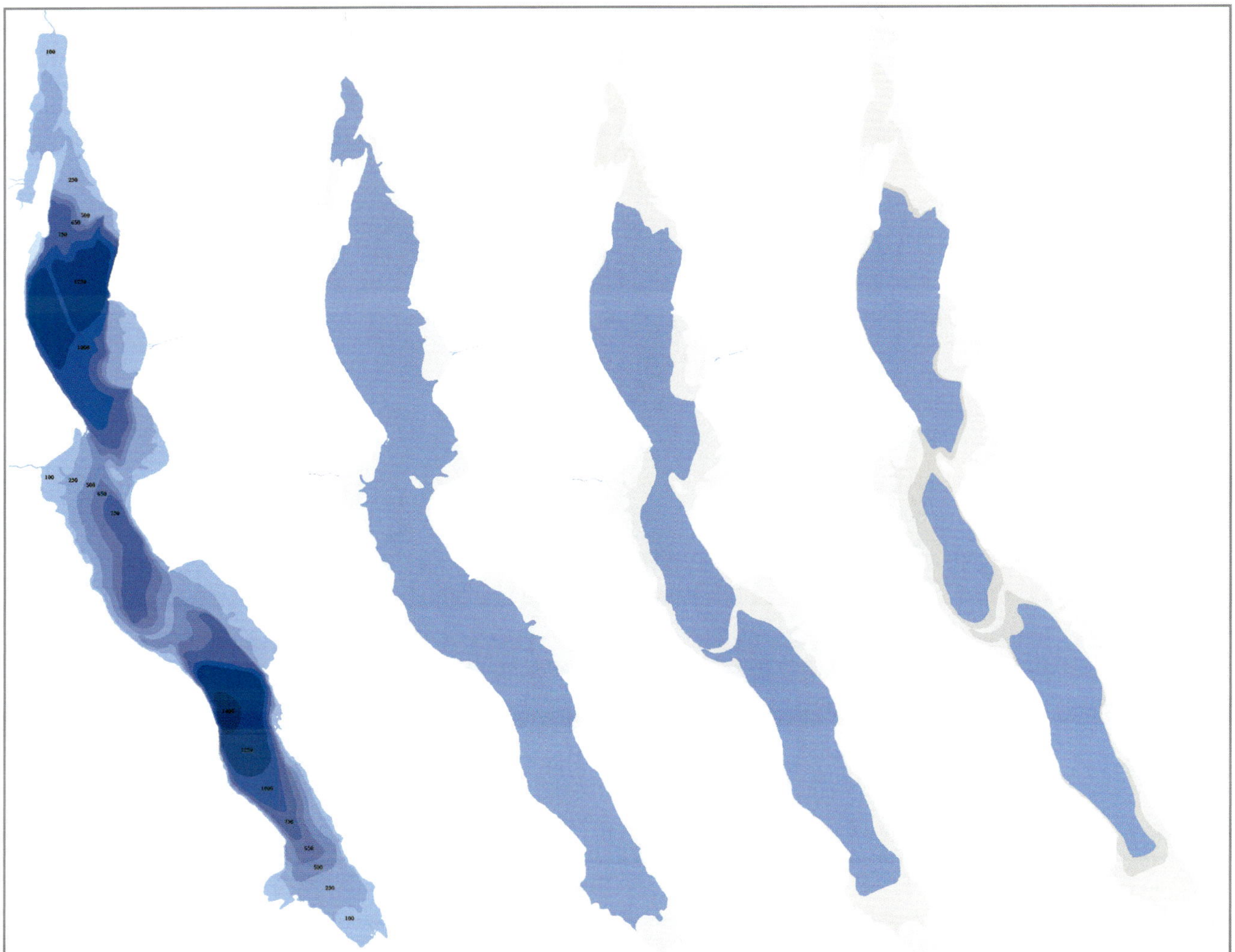

Water level of Lake Tanganyika today	Lake Tanganyika with a water level 100 m lower	Lake Tanganyika with a water level 250 m lower	Lake Tanganyika with a water level 500 m lower

they are likewise still, apparently, the only form, three populations were recently studied, apparently confirming my theories. It has not yet been possible to study the *Tropheus* of the western shore. In 1994 I arranged to *Tropheus* from the central west coast at Yunga sent to me for purposes of study. Unfortunately at the time nobody thought of preserving specimens for research. At that time the collectors were already reluctant to visit this coast, but they did it. Today the situation there is unpredictable, and likely to remain so in the immediate future.

Tropheus duboisi and related forms (lineage 10) represents the oldest surviving lineage. From its current distribution pattern we can deduce that we are looking at relict populations, in particular those at Bemba (in the extreme NW of the distribution of that lineage) and at Karilani Island (in the extreme SE of the distribution of that lineage). The present main distribution extends from Gombe to Kigoma and across the mouth of the Malagarasi to Cape Karago. *Tropheus* of the linage 10 are said to be relatively common along this stretch of coast. Perhaps its original ancestors entered the primeval Tanganyika basin via the Malagarasi, and then, when the lake level rose again, spread from there to the coast at Bemba and Cape Kungwe to the south. From Gombe to Kigoma and thence to Karago lineage 10 is even more common. BRICHARD hoped to find *T. duboisi* on the opposite coast as well, but was unable to do so in the time he had available. For a variety of reasons the coast there is poorly studied and perhaps *Tropheus duboisi* will one day also be found in the vicinity of the Kavala Islands and the Ubwari peninsula.

Because of the long period of time needed for a species range to be reduced, the details of that process cannot always be established. But in all probability it is likely that lineage 3 expanded successfully (and may still be doing so) and slowly forced ancient lineages out of their ancestral distributions. Lineage 5 appears less able to resist younger lineages and is being succumbing to pressure from lineage 3. Apparently precisely the same happened in the case of *Tropheus duboisi*, which has only a very limited range in the north-west. Far to the south in the lake, between Kisambali and Namansi, lineages 8 and 9 intersect, with the fishes of lineage 9 supposedly living below the lineage 9 populated zone on the rocky littoral.

No information is available regarding possible mutual restrictive effects where other populations come into contact.

Today there are still a number of populations, such as the *Tropheus* from Kapampa (pop. 6.1.), all the western populations of lineage 5 from the Congo, some of the north-eastern *Tropheus* of lineage 8 (eg those from Namansi, Hinde B, and Kisambali), which have been little or poorly studied. However, these populations would probably make no great changes to our current knowledge, although they would supplement it further and perhaps reveal one or more additional links.

Map of Lake Tanganyika, showing the localities where *Tropheus* are most often collected.

Tropheus of lineage 1

The *Tropheus* of this lineage can be distinguished from the other lineages in the lake by several characters. The base colour consists of a very dark black-brown. With the exception of the dorsal, all the fins are the base colour. The secondary colours are yellow and red and are expressed as spots or bands on the central flank or the dorsal. With one exception (Caramba 1.9) there is also colour on the head. There are two populations that are completely black (Lueba, pop. 1.51 and Minago, 1.2). Adult specimens never exhibit a stripe pattern. Juveniles lose their stripes very early on compared to other lineages. This lineage has five anal spines. According to a study by Pierre BRICHARD, of 172 specimens collected 170 had five anal spines, one had four, and one had six. In adult specimens the posterior edge of the caudal fin is slightly concave. Depending on mood, the anterior two thirds of the fish may become much lighter, while the posterior third, by contrast, remains black.

This lineage formerly inhabited the northernmost part of the lake, which today is covered by the sediment from the Ruzizi River. In the past the lake is thought to have extended north almost to where Lake Kivu (which did not then exist) lies today. This was supposedly the situation about 10,000 years ago. Geologically speaking this is not a very long period, and many of the current colour variants may have already existed at that time. The rocky biotopes that existed then have subsequently been buried beneath deposits of sediment carried by the Ruzizi river and the populations that lived there have disappeared. These ancient lineages have today been forced further south by the Ruzizi sediment. Many of their biotopes are becoming increasingly silted up, and some of the populations there are now living under poor conditions and their numbers have already significantly decreased. They now survive only in the extreme northeast of the lake as far as Rumonge, and in the west from Mboko to Cape Caramba in the south of the Ubwari peninsula.

At present 10 populations are known. In my view these could possibly be split into two lineages, but the data currently available are insufficient to take this step.

The lineage is sympatric with lineage 5 along the south-eastern side of the Ubwari peninsula. According to BRICHARD, the population densities o a number of populations are decreasing due to severe sedimentation. According to STURMBAUER and MEYER the TROPHEUS from Rutunga and those from Kiriza are more closely related to one another than to the *Tropheus* at the village formerly known as Bemba. A black *Tropheus* has recently been discovered and studied at the village of Kabezi, which lies a few kilometres south of Bujumbura and not far from the Rutunga *Tropheus* population. This has proved to belong to a different, very ancient lineage, and is not related to the black *Tropheus* of lineage 1. The *Tropheus* of lineage 2 are similar to the „black" group in appearance, but are not related to them and derive from another main lineage.

The lineage contains the populations numbered 1.1-1.9..

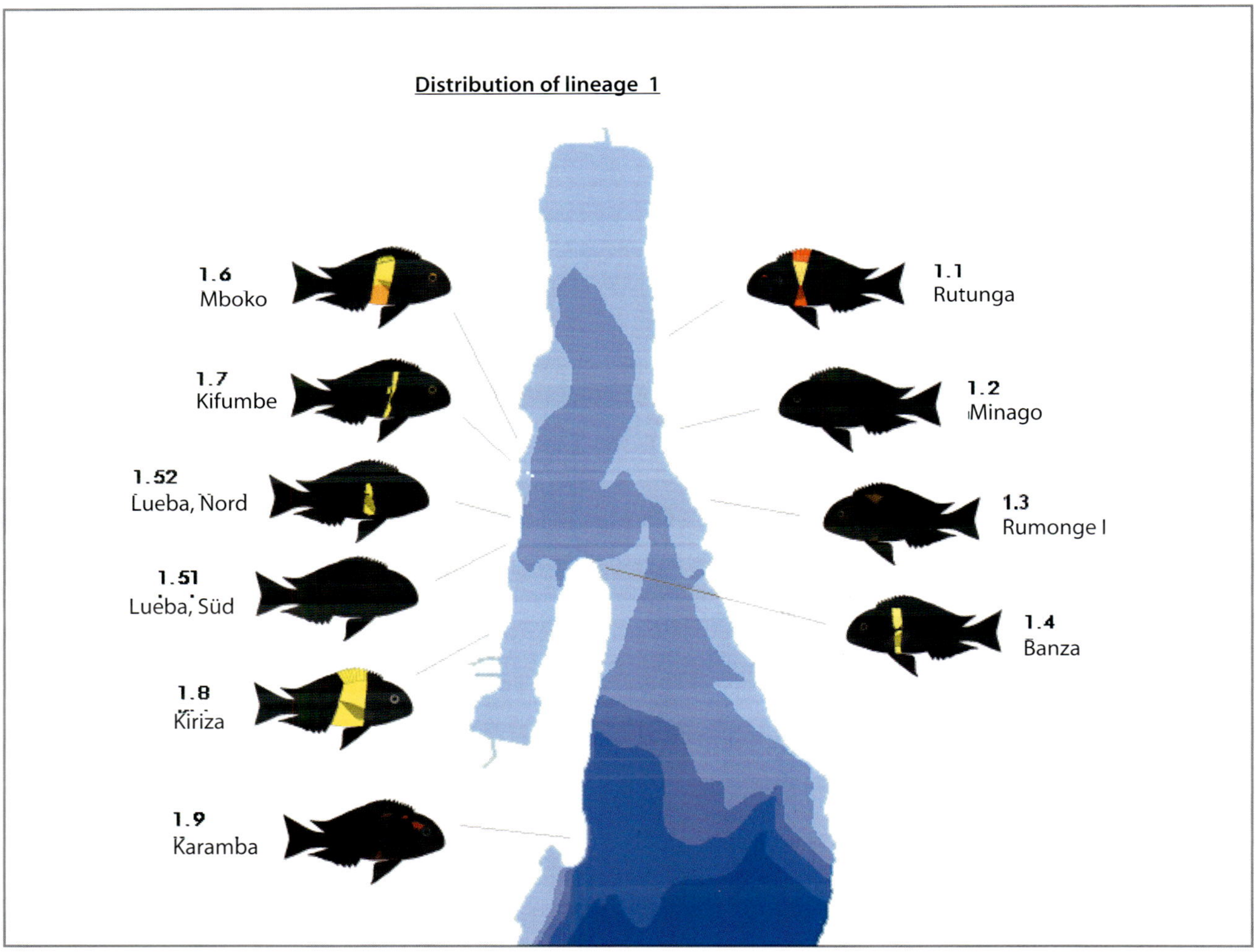

Tropheus sp. Rutunga; Population 1.1

Trade names: Tropheus "Brabant", Tropheus "Rutunga"

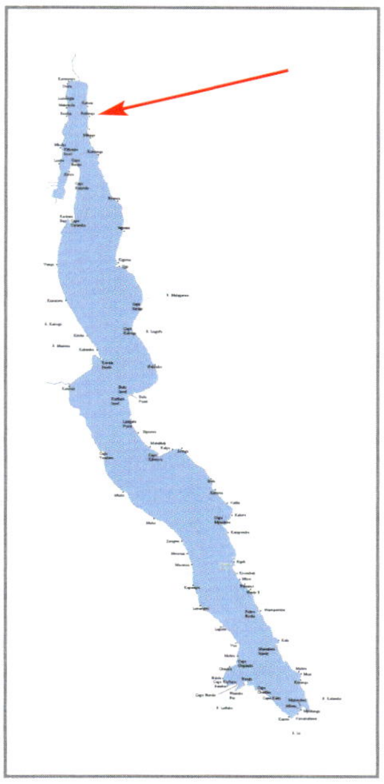

Distribution: South of Kabezi, which lies a few kilometres south of Bujumbura, to Magara. A total of 20 km of coastline, and interrupted only by the mouth of the Mutumba.

Discovered: First mentioned by MARLIER in 1959.

First importation: The first importations were made by Pierre BRICHARD, initially in 1958 and then again in 1971.

Coloration: The base colour is black. The area from between the 5th and 9th (10th) dorsal spines is red. Immediately below there is a golden yellow inverted triangle with a red reticulated pattern. Immediately below this there is a red triangle whose apex meets that of the upper, inverted, triangle. Depending on mood, behind these triangles there may be a narrow lighter area, in which the bases of the scales are red. Again depending on mood, the black colour of the anterior two thirds of the body may pale to light grey. The black colour of the posterior third, and the coloured triangles, remain unchanged. All the fins are likewise black. The upper edge of the eye is dark red.

Remarks: his colour form has long been known as the "Brabant" and for many years was the only form imported in any numbers. Unfortunately it is now found only in the tanks of enthusiasts, as its popularity has waned as numerous other forms have become available. This is difficult to understand, as well-maintained groups are very attractive, with their high component of red and yellow. Adult males can attain a length of 16 cm.

Semi-adult male with numerous red scales on the body. The black of the posterior body and the red upper edge of the eye are also clearly visible.
photo: P. Schupke

Tropheus sp. Minago; Population 1.2

Trade names: Tropheus "Minago", Tropheus "Black"

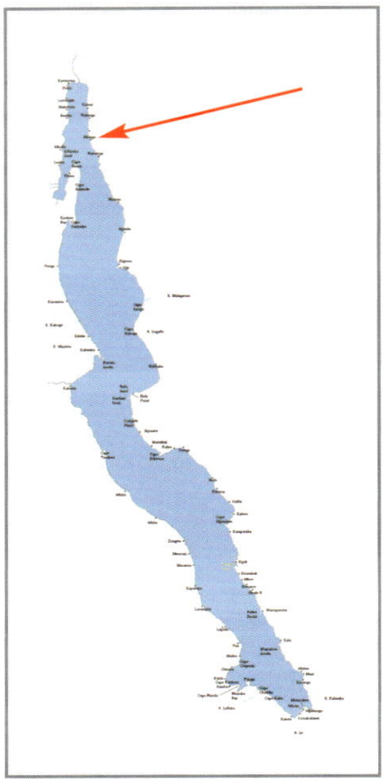

Distribution: Several populations along a 10 km stretch of coast south of the village of Minago in Burundi.

First importation: Unknown. This colour variant has been imported only very rarely and then only in very small numbers.

Coloration: This colour form is completely black all over the body, except that the belly region may exhibit a slight bluish shimmer and there may be a barely perceptible yellowish tinge to the flanks. There are sometimes lighter, washed out areas of colour on the belly and back. These Tropheus are very reminiscent of the Uvira population in the extreme north-west of the Congo.

Remarks: As stated by Brichard, the survival of this population is increasingly more seriously threatened by the silting of its biotope. There are supposedly large but scattered populations at Cape Magara. Because of its inconspicuous coloration this fish has stimulated little interest among aquarists. It is likely that there are no longer *Tropheus* of this colour variant in our aquaria today.

Tropheus sp. Rumonge I; Population 1.3

Trade names: No trade name known for this population.

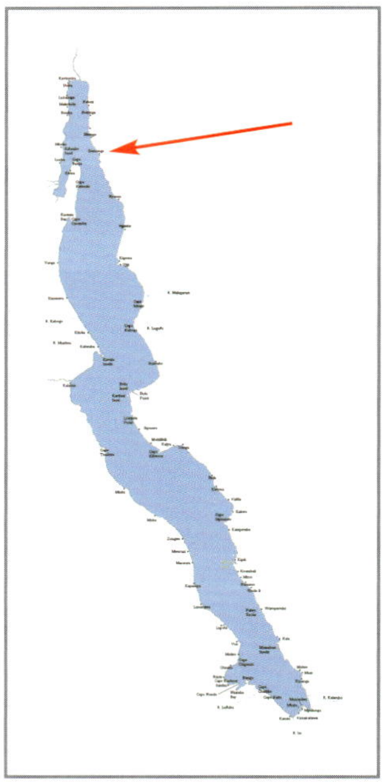

Discovered: Matthes was the first to mention this *Tropheus*, back in 1962.

First importation: No known records of commercial importation.

Distribution: Matthes (1962) cites as locality a stretch of coast a few kilometres north of the village of Rumonge on the central east coast Burundi.

Coloration: According to Brichard, the base colour is black-brown. There is a faint brown area on the anterior part of the back, and immediately above in the dorsal fin, and a beige-brown area on the belly. The iris is silvery-white with a yellow upper edge. The anterior two thirds of this *Tropheus* may, depending on mood, pale to beige-brown. This lightening of the anterior body is characteristic of *Tropheus* of lineage 1; it is particularly noticeable in the *Tropheus* from Rutunga and Cape Caramba, but is also seen in those from Bemba, which do not belong to this lineage.

Remarks: Since 1962 when Matthes first mentions this form, they have never been collected again. In 1984 Pierre Brichard undertook a collecting trip to this region, with a view to comparing this Tropheus with the very similar „Chocolate Moorii" (*Tropheus brichardi*) found not very far away. He thus came to the conclusion that the two *Tropheus* in question must be different species. This population is separated from the „Chocolate Moorii" at Nyanza Lac by a stretch of sandy shore many kilometres in extent. Brichard states that the number of anal spines is generally 5, as in the *Tropheus* of lineage 1, while *Tropheus brichardi* has 6 anal spines. In stress situations these *Tropheus* exhibit the same lightening of the anterior body as is seen in other individuals of lineage 1. According to Konings, the same stretch of coastline is home to a *Tropheus* with a green-yellow band, which apparently belongs to lineage 5 (Pop. 5.14).

Tropeus sp. Banza; Population 1.4

Trade names: Tropheus "Banza"

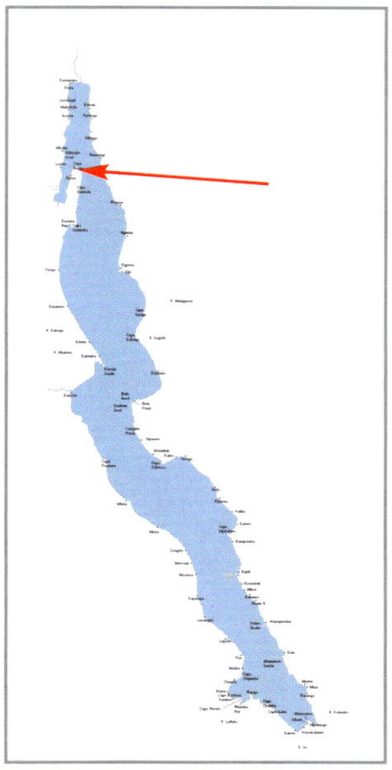

Distribution: From the village of Manga, which lies somewhat to the south-west of the northern tip of the Ubwari peninsula, to Cape Muzimo, a stretch of coastline about 50 km long.

First importation: Unfortunately not imported to date.

Coloration: The body is completely black apart from a narrow lemon yellow stripe, which may extend from the belly to the dorsal or be restricted to the upper dorsum. The band is sometimes interrupted centrally. The upper half of the eye may be brilliant yellow in colour.

Remarks: This *Tropheus* resembles the Kiriza form (1.8) very closely. This may reflect the immediate proximity of the Kiriza variant. Perhaps these two forms constituted a single population at some time in the past when the lake level was lower, and became separated as the level rose again. Like the Lueba form below, this form is very variable in its coloration.

Tropheus sp. Lueba; Populations 1.51 and 1.52

Trade names: None known

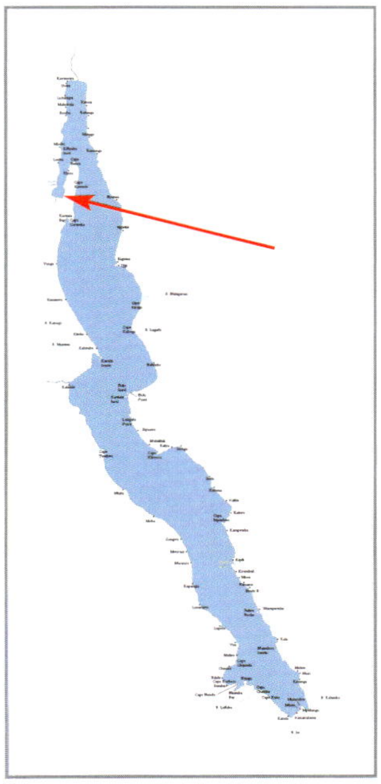

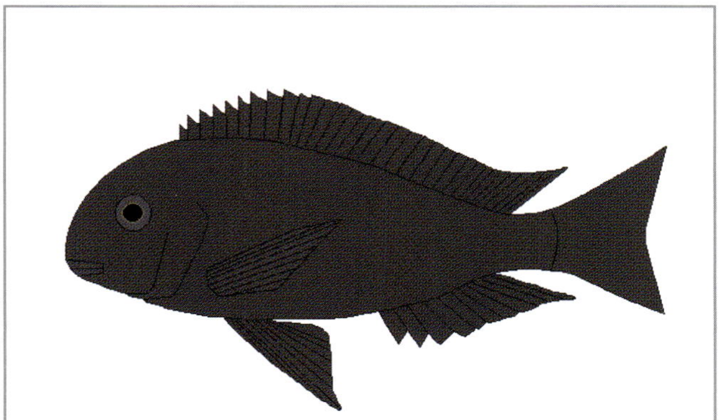

1.51

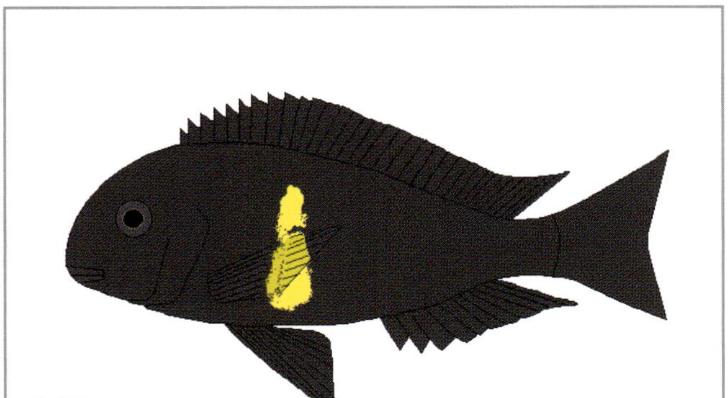

1.52

Distribution: These *Tropheus* live in separate areas about 3 km apart.

First importation: This colour form has probably never been imported.

Coloration: The northern population is more variable than the southern and may have a „teardrop" spot on the side. Some individuals may have two additional spots on the flank. Many specimens are completely black, including, supposedly, those from the immediate vicinity of Lueba.

Remarks: These *Tropheus*, which were discovered by Matthes and Marlier, are divided into two populations which are separated by a stretch of heavily sedimented rocky coast about 3 km long. The northern population contacts with the Mboko population a few kilometres further north. Brichard is of the opinion that the future survival of this form is in doubt, so it would be wise to collect a few hand-picked specimens in order to breed them for the aquarium hobby.

Tropheus sp. Mboko; Population 1.6

Trade names: Orange-Moorii I

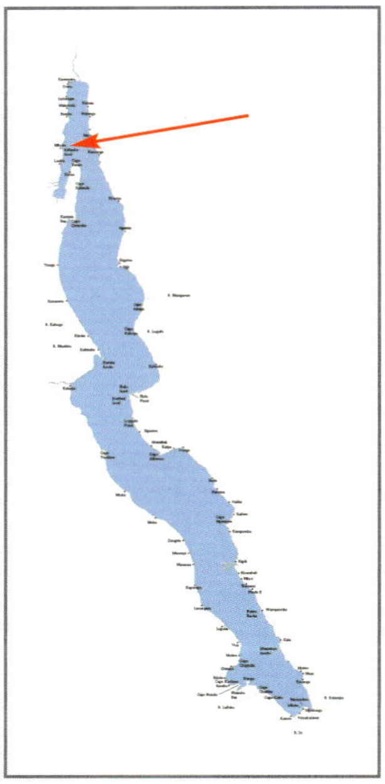

Distribution: The locality for this population lies a few kilometres south of Cap Munene (Bemba), to the north of the village of Mboko.

Coloration: The base colour is black-brown. In this *Tropheus* the band is up to 50% narrower than in specimens from Bemba. The upper part of the band is almost lemon yellow, while the lower half recalls the orange of the Bemba population. Often the band does not extend into the dorsal. The eye is dark apart from a barely visible brown-red upper edge.

Remarks: *Tropheus* sp. "Mboko" can be confused with *Tropheus* sp. "Kiriza" (1.8), which has a slight orange reticulation on the lower half of its lemon-yellow band. The *Tropheus* sp. "Kiriza" from the southernmost part of their range have a completely yellow band in the dorsal, while *Tropheus* sp. "Mboko" does not. *Tropheus* sp. "Kiriza" are only rarely imported.

Tropheus sp. Kifumbwe; Population 1.7

Trade names: Tropheus "Lemon Stripe"

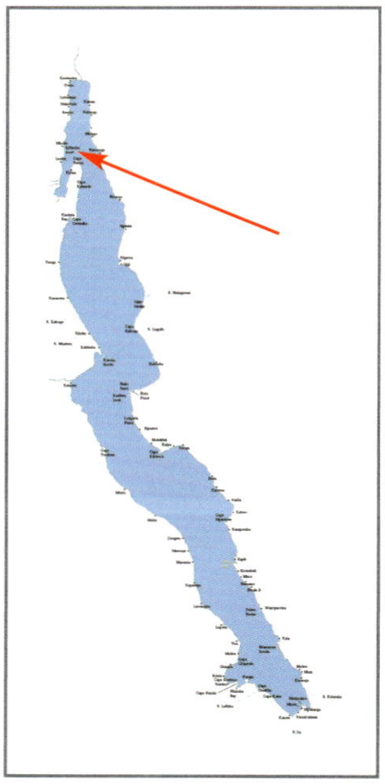

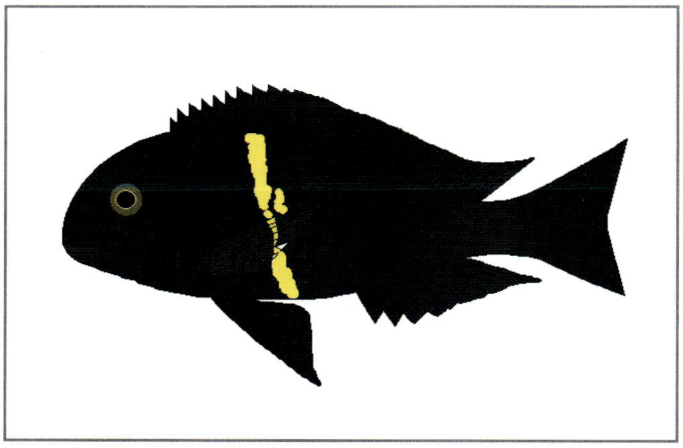

Distribution: This form inhabits a number of islands offshore of the village of Mboko. There is an underwater link with the mainland, where scattered *Tropheus* are also found.

First importation: The first consignments reached Europe at the beginning of the 1980s.

Coloration: Apart from a narrow lemon yellow transverse stripe, this colour form is completely black. The upper part of the eye-ring is dark red to some degree. The yellow band is restricted to the body, and the dorsal is plain black. The throat and mouth region may exhibit a slight blue sheen.

Remarks: Brichard estimated the number of wild *Tropheus* of this variant at no more than a few thousand individuals. This *Tropheus* was bred for the trade in ponds at his station in Burundi. Unfortunately demand is low, and hence this colour form is correspondingly uncommon in our aquaria.

Tropheus sp. Kiriza; Population 1.8

Trade names: Tropheus "Kiriza", Yellow Kaiser

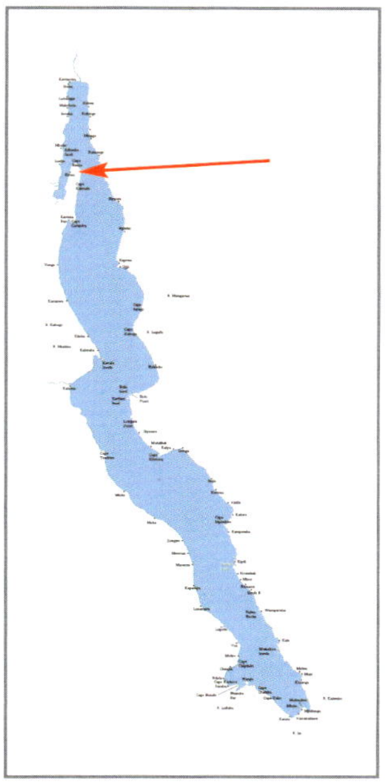

Distribution: A 35 km stretch of the west coast, southwards from near to the northern tip of the Ubwari peninsula.

First importation: Supposedly in the mid-1980s.

Coloration: The black base colour is interrupted by a broad yellow-orange band which extends into the dorsal. Males have a rather broader band than females. There are slight colour variations within the relatively large distribution, noticeable in the intensity of the band. Individuals in the southern part of the range have an orange reticulation in the lower part of the band. Depending on mood, the black anterior body may become much lighter, turning light grey. The band may become narrower, again depending on mood. The upper outer edge of the eye is rather orange; the remainder is dark.

Remarks: According to the researches of STURMBAUER and MEYER, the *Tropheus* sp. from Kiriza and that from Rutunga (1.1) are closely related and belong to an ancient lineage. They are supposedly more closely related to each other than to the *Tropheus* sp. from Bemba (2.4). This very attractive colour form was bred in ponds for the aquarium trade by BRICHARD´s company in Burundi. The captive-bred individuals did not always achieve the size and quality desired. In the mid 1990s large numbers of these *Tropheus* were collected and exported by CJ Aquarium. The colour intensity of the band was almost unsurpassable. They are difficult to keep in pairs because of the aggression of the males, but there are fewer problems if they are kept in a larger group.

Adult male in full colour, showing the broad band. **photo:** P. Schupke

Tropheus sp. Caramba; Population 1.9

Trade names: Tropheus "Caramba", Tropheus "Katende".

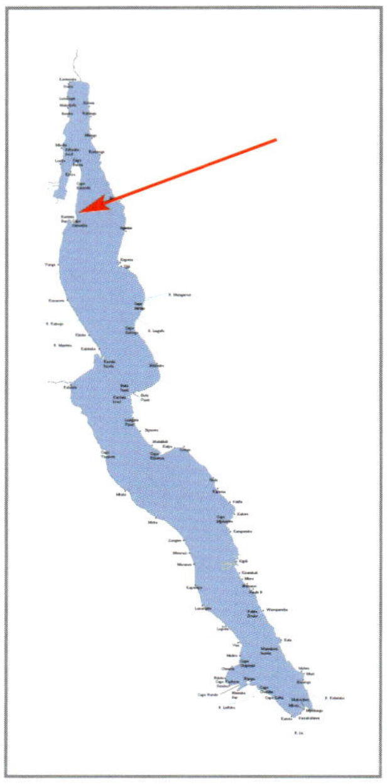

Distribution: The distribution begins at Cape Muzimo and extends as far as Cape Caramba at the southern tip of the Ubwari peninsula, a stretch of coastline about 50 km long.

First importation: The first commercial importations were at the beginning of the 1980s, and came from the company "Fishes of Burundi", headed by Pierre Brichard.

Coloration: The base coloration, including fins, is black-brown. The posterior part of the head, the cheeks, the dorsum below the 5th to 9th dorsal spines and the base of the latter, exhibit a red colour which may appear lighter or darker depending on the collecting locality. In a few individuals there may be two indistinct triangles with their apices meeting on the centre of the flank. This pattern is very similar to that of *Tropheus* sp. "Rutunga" (1.1). The upper edge of the eye is an attractive red, and the rest of the iris may have a light blue sheen. The lower part of the operculum may exhibit a turquoise hue. Juveniles have significantly larger red areas corresponding to the coloured markings in adults.

Remarks: Unfortunately, these *Tropheus* are rarely exported. They can be very intolerant of one another. They are bred in ponds in Burundi for the aquarium trade. They are very similar to *Tropheus* from Rutunga (1.1) in that some individuals exhibit vague coloured triangles. These triangles are more clearly visible in juveniles of the Rutunga *Tropheus*. This similarity undoubtedly indicates a close relationship in terms of the age of this population. Along the east coast of the Ubwari peninsula *Tropheus* sp. "Caramba" lives sympatric with a *Tropheus* of lineage 5 (Pop. 5.4), whose population density is greatest at the central point of the peninsula and decreases further to the south, where *Tropheus* sp. „Caramba" is said to attain a higher population density.

Tropheus of lineage 2

Lineage 2 includes *Tropheus brichardi,* one of the few described *Tropheus* species. Unfortunately *Tropheus brichardi* was described at a time when many other populations remained unknown. The genetic sequencing of 54 individuals of 21 populations by Sturmbauer and Meyer in 1990 produced a surprising result regarded as incomprehensible in some circles: in their view *Tropheus brichardi* is genetically identical with the *Tropheus* from Bemba. Whether or not one now accepts the results of this research is essentially unimportant. But even an antipathy to this type of research does not permit its results to be completely swept under the carpet. In my view, however, the available evidence does not permit the brown *Tropheus* from Nyanza Lac (*Tropheus brichardi*) and further south at Gombe to be assigned to any of the other lineages. Moreover, it seems to me that the *Tropheus* from Uvira (2.3, not those of lineage 5 that also occur there), the *Tropheus* from Makobola (2.5, not lineage 5), and the *Tropheus* from Bemba (2.4) do not fit into lineage 1 (the black lineage), as they exhibit a somewhat different external appearance. The two lineages 1 and 2 differ in a number of details in their external appearance. The base colour of lineage usually consists of dark brown rather than black as in lineage 1.

The known distribution of lineage 2 extends from north of Nyanza Lac to Ngombe, a few kilometres further south, and in the north-west from Uvira to Bemba. According to Brichard, the Mboko population contacts the Bemba population. According to the results of DNA research, lineage 2 evolved after lineage 1, and is thus phylogenetically younger. The "brown" lineage may have originated in the area to the north of Kigoma.

Recent research justifies assigning these populations to a lineage of their own. It is true that there is an external resemblance to lineage 1, but we must not forget that a degree of correspondence in coloration can arise from parallels in evolutionary pressures such as predation, water conditions, and other biotope factors. The past variation in the water level of Lake Tanganyika makes precise study of this puzzle far more difficult.

This lineage contains the populations 2.1 - 2.5.

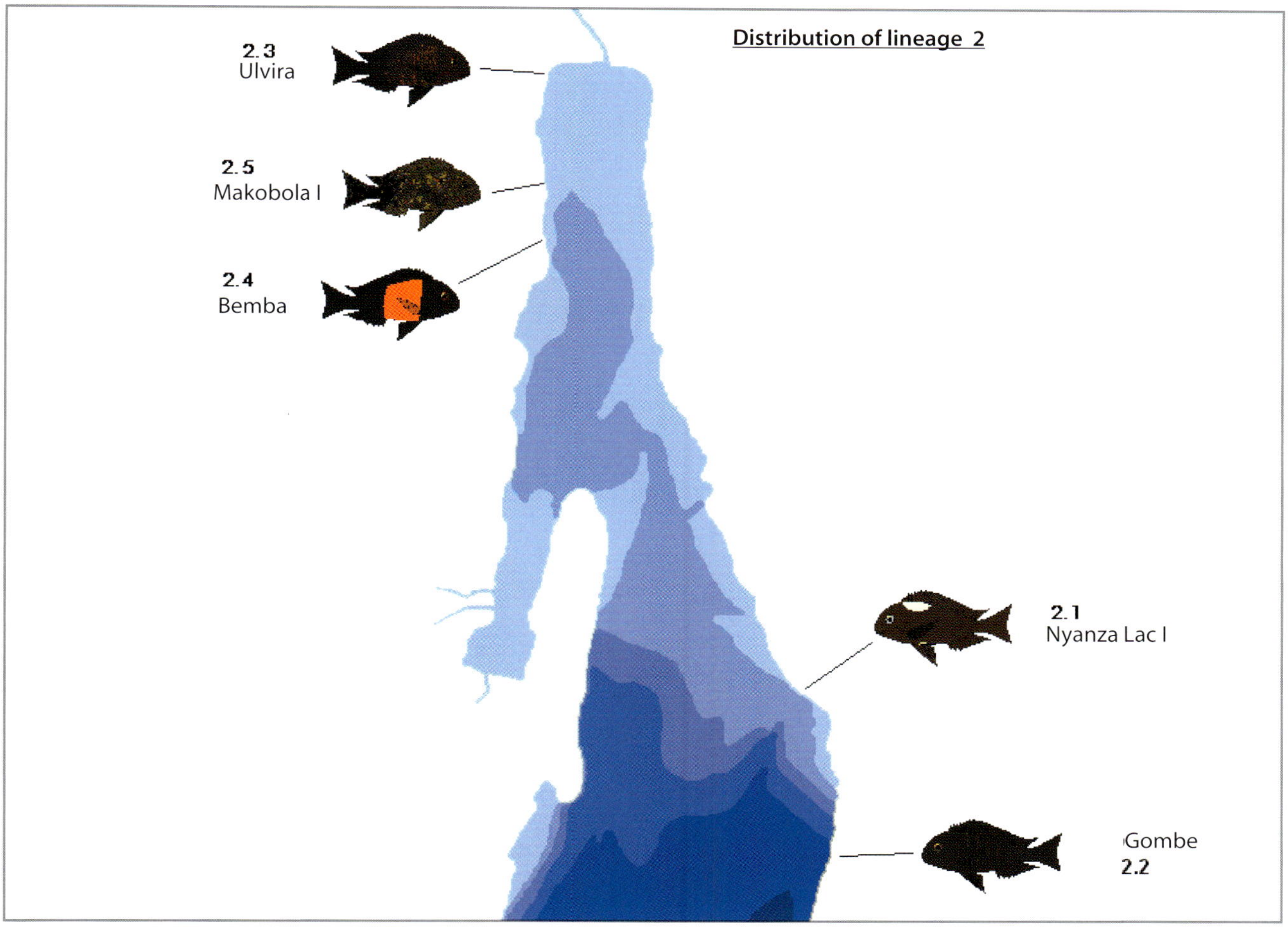

Tropheus brichardi NELISSEN & THYS, 1975; Population 2.1

Trade names: Chocolate Moorii, Tropheus "Chocolate", Tropheus "Brown", Tropheus "Nyanza Lac I"

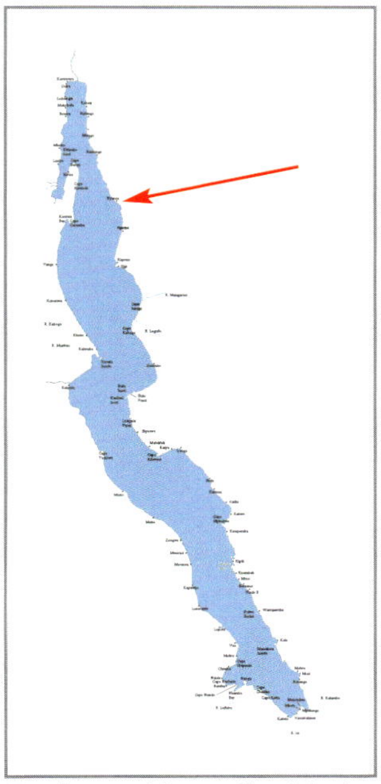

Distribution: The known distribution lies a few kilometres north of Nyanza Lac and extends for about ten kilometres of coastline.

First importation: The Chocolate Moorii has been exported since 1972 by the company Fishes of Burundi.

Coloration: The entire fish is dark brown with a whitish-yellow saddle spot located beneath the fourth to thirteenth dorsal spines. There may also be a smaller yellowish spot on the anterior part of the belly. The upper part of the iris is yellow in colour. Juveniles have whitish-yellow transverse bars on a dark brown body.

Remarks: According to statements made by BRICHARD, inter alia, there are three differently coloured *Tropheus* populations at the coast at Nyanza Lac, although BRICHARD actually speaks of a high variability in the *Tropheus brichardi* living there. The same applies to the coast at Rumonge . In this area there are not only the regular chocolate *Tropheus* (lineage 2) but also individuals with green-brown stripes and a blue iris (lineage 5, Pop. 5.13) and a brown/whitish-yellow striped *Tropheus* with a yellowish dorsal (lineage 3). It is not always easy to assign *Tropheus* in the lake to a specific population quickly and accurately, as the fishes are well adapted to the ambient water conditions and local background. At Ngombe, which lies about halfway between Nyanza Lac and Kigoma, there is a *Tropheus* which belongs genetically to lineage 2.

Tropheus sp. Gombe; Population 2.2

Trade names: No trade names known for this population.

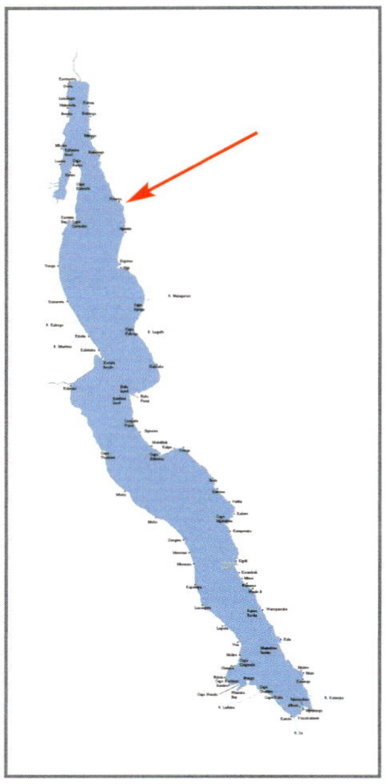

Distribution: The distribution of this *Tropheus* lies a few kilometres south of Nyanza Lac north of the Tanzanian border. It is found close to the surface in suitably rocky biotopes.

Discovered: I can find no records in the literature as to when this form was first discovered.

First importation: This population has not yet been imported for the aquarium trade, but has found its way to Europe in the form of research material preserved in alcohol.

Coloration: The entire fish is dark brown on the body, including the fins. A barely perceptible transverse barring is visible in the fish in the underwater photo, but it is not possible to tell from the photo whether the fish is male or female, or what mood it is in. There is no sign of a spot on the back that would indicate coloration close to that of *T. brichardi*. The dark iris is rather yellower above than below.

Remarks: This *Tropheus* population was discovered only a few years ago. I saw the first photos, taken by Patrick Louisy, in 1994. The external similarity to *T. brichardi* from Nyanza Lac is obvious, and the „internal" relationship has been shown to be equally unambiguous by genetic research. The same stretch of shoreline is home to *T. duboisi*, here in the northernmost part of its east coast range. Nowadays a river mouth with kilometres of sandy shore clearly represents an insuperable barrier to movement between Gombe and Nyanza Lac. It is not known whether other *Tropheus* lineages inhabit this stretch of coast.

Tropheus sp. Uvira; Population 2.3

Trade names: None known.

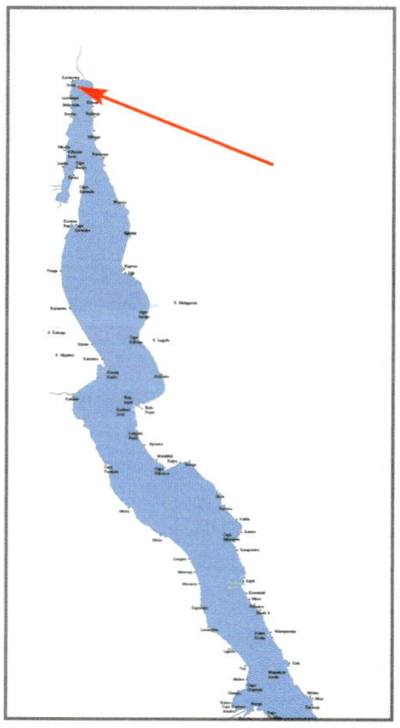

Distribution: The distribution extends from the extreme north-west of Lake Tanganyika at Kavimvira, past Uvira to the mouth of the Kivovo.

First importation: According to Brichard, this form has not been exported because of its dull coloration.

Coloration: The majority of the population are brown-black in colour. A few specimens have an incomplete dark orange band on the sides, but generally they are sooty black. The upper half of the eye is brown.

Remarks: This population is split between two areas, one in the region of Uvira, the other in the extreme north-west at Kavimvira, where these fishes are sympatric with Tropheus that probably belong to lineage 5 (Pop. 5.1).

Tropheus sp. Bemba; Population 2.4

Trade names: Tropheus "Bemba", Tropheus "Orange Band", Orange-Band Moorii

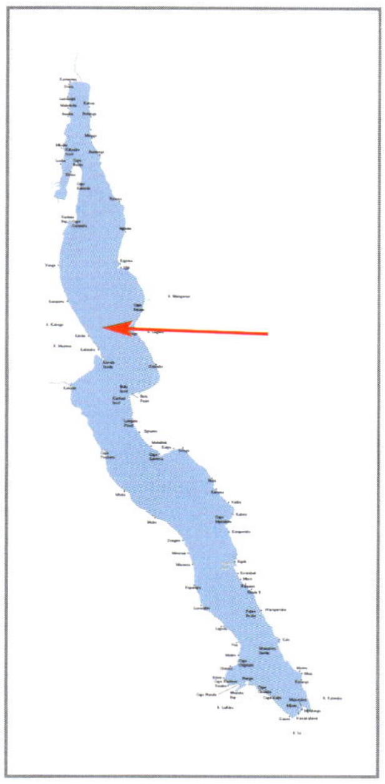

Distribution: The distribution of this population encompasses about a kilometre of shoreline at Cap Munene on the north-western coast of the Congo.

Discovered: This *Tropheus* was first mentioned by Marlier as long ago as 1959.

First importation: The first specimens reached Europe back in 1958. Further importations did not take place until 1976.

Coloration: The brown-black base colour is interrupted by a bright red-orange band, very variable in its width, on the central flank and extending to the upper edge of the dorsal. Depending on mood, the anterior third of the body may become much lighter and assume an orange hue. According to Brichard there are individuals in which the anterior part is red-orange. The posterior third, by contrast, is always dark. The upper edge of the eye is also orange. Juvenile specimens have light transverse bands, which they start to exhibit at a length of only a few centimetres.

Remarks: This *Tropheus* population is sympatric with *Tropheus duboisi* at Cap Munene. *Tropheus* sp. "Bemba" was no longer imported for many years, and regular consignments of this attractive fish did not resume until the beginning of the 1990s. The stocks imported at that time had a particularly brilliant cinnabar-red band and captivated many aquarists. I purchased 40 fry from the first tank-bred generation and placed them in a 1,200 litre tank until they were sexually mature. They were a real sight for sore eyes. Although the parents exhibited exceptionally broad bands, the youngsters had numerous narrower, incomplete bars, and colour intensity was variable too. In the native biotope too, not all individuals have permanent broad bright bars. These fishes presented an interesting picture in the morning when they first awoke: some individuals appeared to have lost their band, while others, by contrast, had become orange on the head. After a few minutes, however, the melanin in the skin had returned to its proper place. According to Robert Allgayer, the correct name for the collecting site is "Pemba".

Adult, sexually active male of Tropheus sp. "Bemba". **photo:** P. Schupke

Tropheus sp. Makobola I; Population 2.5

Trade names: There are no trade names for this population.

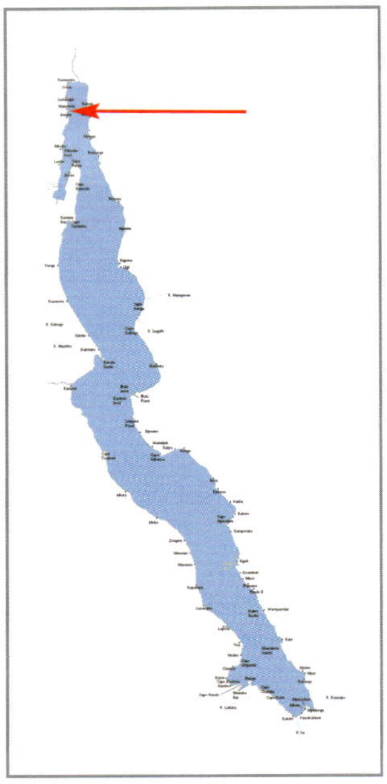

Distribution: The distribution of this *Tropheus* begins at Makobola and ends shortly before the mouth of the Kivovo.

Coloration: The markings in this local form are very variable, ranging from narrow, pale orange bands to copper-orange spots on a black background, or even completely black.

Remarks: The *Tropheus* of this section of coast are not geographically separated from the range of *Tropheus* sp. "Bemba" (2.4), and, according to BRICHARD, in the north they are still in contact with the Uvira population (2.3) although the mouth of the Kivovo constitutes an increasingly effective barrier. The former Makobola-Luhanga population has probably been invaded by the Bemba and Uvira populations, and now exhibits a high degree of variability. At Makobola also a *Tropheus* of lineage 5 (Pop. 5.2) exists, which is catalogued here as *T.* sp. Mabokola II.

Tropheus of lineage 3

This *Tropheus* lineage occupies a 200 kilometre stretch of coastline in the former Zaire, beginning in the north at Zinzia Creek (halfway between the southern end of Ubwari peninsula and the Kavala Islands to the north of Kalemie) and ending - according to the latest data - at M´toto on the central Congo coast. In Tanzania the distribution of this lineage begins in the south on the mainland at Karilani Island, and - according to unconfirmed reports - extends even further south. The rocky coast does continue further southwards. The northern limit to the range is nowadays at Nyanza Lac in southern Burundi. Lineage 3 lives sympatric with several other *Tropheus* lineages on the Tanzanian coast: these are the *Tropheus* of lineage 10, the blue-eyed greens of lineage 5, the fork-tailed *Tropheus* of lineage 4, and *Tropheus* sp. "Kirschfleck" of lineage 12.

In both Tanzania and the former Zaire the base colour tends to be grey or grey-brown in the northern part of the range and browner in the south. The fins, apart from the pectorals and caudal, are more or less yellow, yellow-brown, or brown in colour. The soft parts of the dorsal and anal often have a blue to turquoise metallic sheen. Juveniles and females have yellowish-white barring all over the body. Dominant individuals lose their stripes, but may resume them if circumstances change. All fishes of lineage 3 have a yellow, sometimes orange-yellow, hue on the head or body. They have 6 anal spines. The upper part of the eye is yellow, yellowish, or orange-yellow in colour. All share the common feature of becoming much darker when sexually active, and this coloration may be permanent in powerful males occupying territories.

Little information is available regarding local forms south of Kalemie. There are more than 100 kilometres of inadequately explored coastline between Kalemie and M´toto. Only at Cape Tembwe, approximately in the middle of this stretch of coast, has a further local form of lineage 3, uniform brown in colour, come to light. According to BRICHARD rocky shoreline alternates with sandy bays along this coastline. This region must, without doubt, harbour several more populations of lineage 3, similar *Tropheus* sp. "Kalemie", *Tropheus* sp. "M´toto" and the *Tropheus* sp. from Cape Tembwe. Perhaps, however, there will also prove to be an as yet undiscovered *Tropheus* of an older lineage in this region, reduced to a small relict population or forced into deeper water by pressure from lineage 3. Lineage 3 appears to be relatively recent and may well be still be in the process of colonising suitable rocky coastline at the expense of other *Tropheus* lineages. Lineage 3 is found sympatric with several other lineages within its large distribution: at Nyanza Lac in the north with lineages 2 and 5; at Kigoma with a *Tropheus*, which belongs to lineage 10. In the most south-easterly part of its range, at Bulu Point at the foot of the Kungwe Mountains, it occurs with *Tropheus polli*, which belongs to lineage 4, and *Tropheus* sp. „Kirschfleck" from lineage 12.

In the south-west, between Kunanwa and Toa, it is sympatric with the blue-eyed greens of lineage 5, and with *Tropheus* sp. "Kongole", which is very similar to *Tropheus polli* and belongs to lineage 4. Further south, in the area around Pala and M´toto, it meets with a further fork-tailed member of lineage 4. In the past, when Lake Tanganyika was divided into three lakes, lineage 3 occupied a coastline in the southern end of the northern lake .

This lineage contains the populations numbered 3.1-3.15.

Distribution of lineage 3

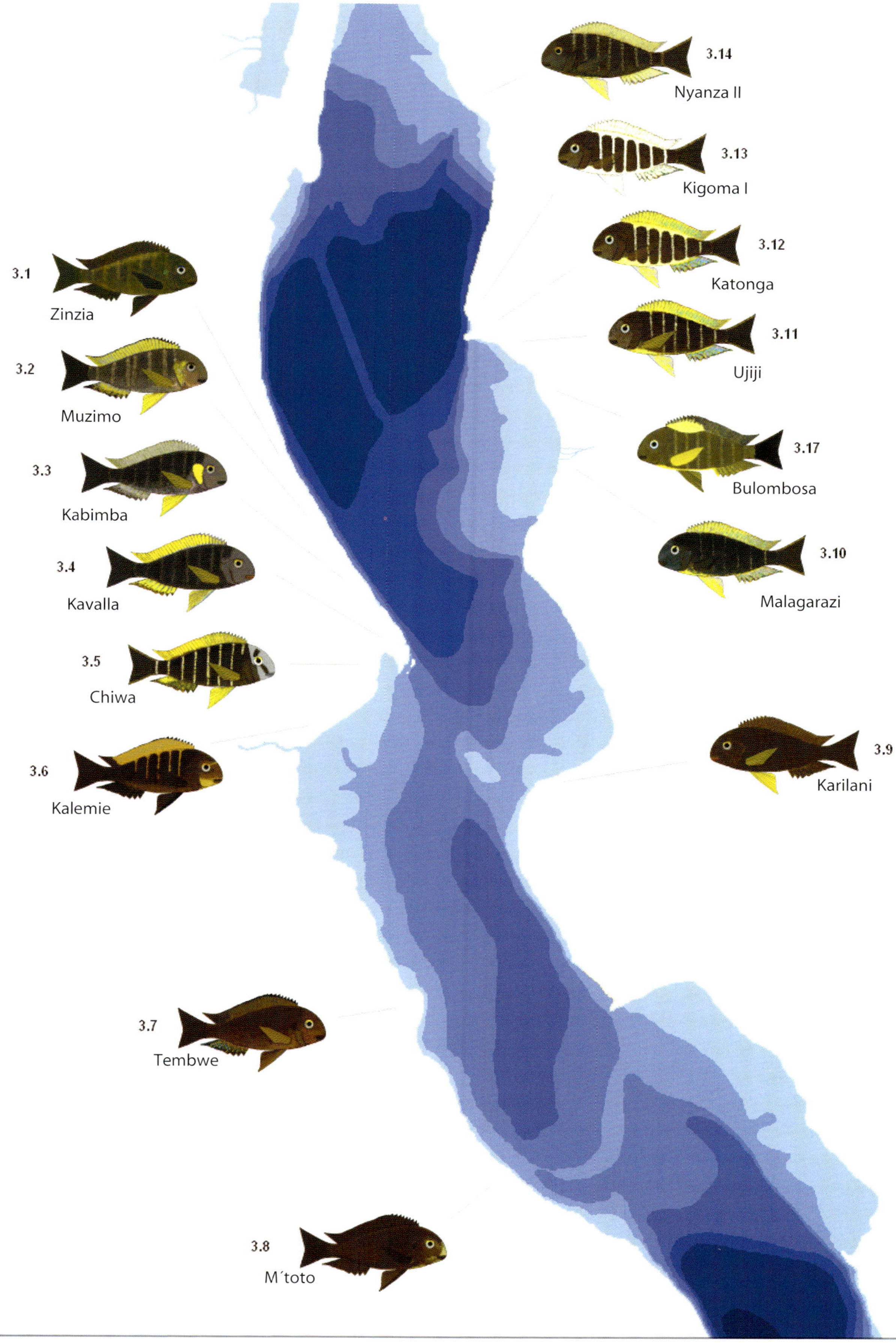

Tropheus sp. Zinzia; Population 3.1

Trade name: These Tropheus have not yet been imported.

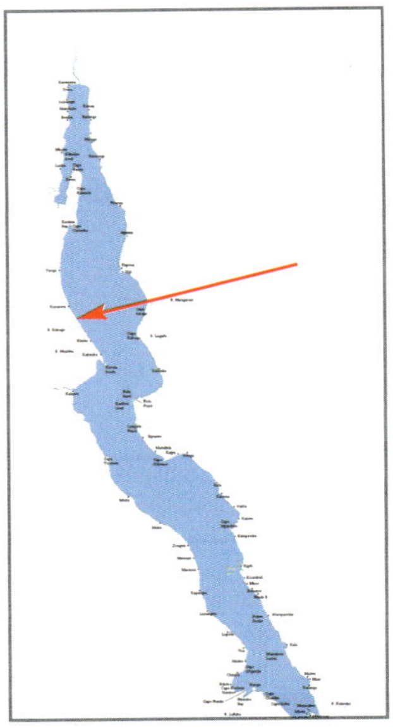

Distribution: The distribution of this *Tropheus* begins to the north of the Kabogo river and Zinzia Creek.

Discovered: Pierre BRICHARD and his collecting team discovered this population during one of their many expeditions around the lake.

Coloration: BRICHARD speaks of a dark green base colour with a faint but large yellow spot beneath the dorsal. All the fins are olive green. The drawing shows much fainter markings than in the following populations of this lineage.

Remarks: This population is rather reminiscent of lineage 5 by virtue of its base colour. Unfortunately no photos of this local form are available. The drawing presented here was made from the description given by Pierre BRICHARD and does not claim to be an accurate representation of the population at Zinzia Creek. *Tropheus* of lineage 5 also live within the distribution of this form.

Tropheus sp. Kabogo; Population 3.2

Trade name: No trade name exists for this local form.

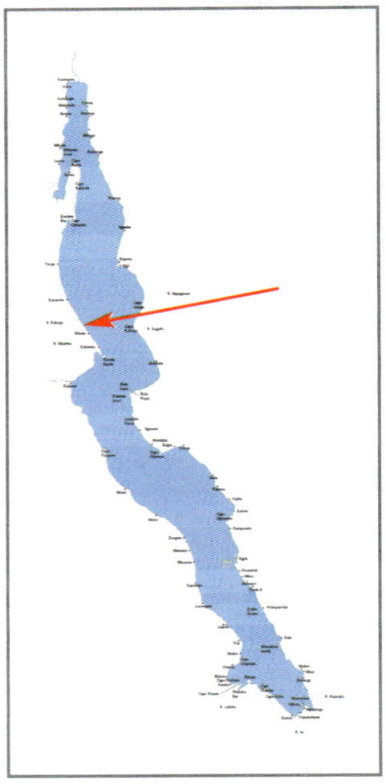

Distribution: This population inhabits an approximately 20 kilometre section of coastline between the Kabogo and Muzimo rivers. It adjoins the previous population at the southern end of the latter´s range.

Discovered : This *Tropheus* was likewise discovered by Pierre BRICHARD and his dive team.

First importation: : This *Tropheus* showed up at a German importer´s at the beginning of the 1990´s.

Coloration: The base colour consists of a dull mouse-grey which appears to have a pinkish overlay and is banded with whitish vertical bars. With the exception of the caudal, all the fins are yellowish, with the pectorals and ventrals a more intense yellow. The dorsal has a black edging. There is a very weakly expressed cheek spot, reminiscent of the adjacent population (3.3) which has a bright yellow cheek spot. The lower part of the eye is lighter in colour than the upper.

Remarks: At the beginning of the 1990s the author discovered a few individuals of the Kabogo form in an importation of *Tropheus* sp. „Kabimba" (the „Canary Cheek", 3.3). Perhaps these *Tropheus* were sent as samples, or perhaps this was a case of a neighbouring population being caught, for whatever reason, with other *Tropheus* as the „Canary Cheek". It would be interesting to find out how neighbouring populations come to exhibit such different colour patterns. Undoubtedly the nature of the rocky littoral, with differently coloured rocks and Aufwuchs, plays a part, as does pressure from predators above and below water. But these factors are unlikely to differ greatly between neighbouring populations. In spite of their striking colours in the aquarium, the different *Tropheus* populations are rather inconspicuous in their native biotopes.

Tropheus sp. Kabimba; Population 3.3

Trade name: Tropheus "Kabimba", Tropheus "Canary Cheek".

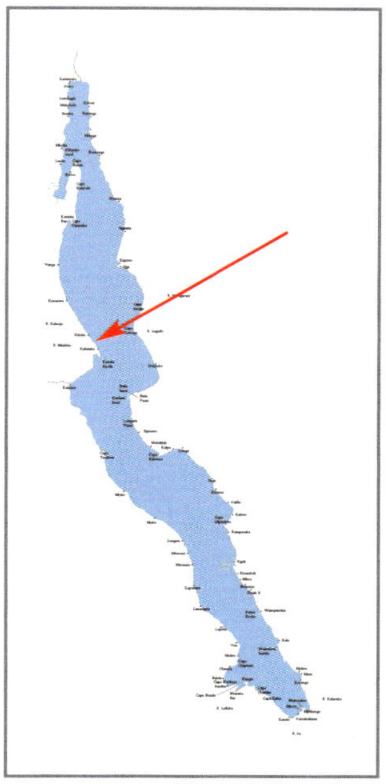

Distribution: The location for this local form lies on an approximately 10 kilometre stretch of coastline between the Muzimo river and Kabimba Bay.

Discovered: This attractive *Tropheus* was discovered by Pierre BRICHARD and his dive team.

First importation: The first imports of these cichlids appeared in the trade at the beginning of the 1990s.

Coloration: The base colour consists of a dark mouse grey with light vertical bars. Depending on mood, the stripes and head may exhibit the same degree of light coloration. Dorsal, pectorals, and anal are a plain yellowish-grey, but the ventrals, by contrast, are bright lemon yellow. The most striking feature is the bright yellow spot on the posterior gill-cover, resembling a tear-drop on the side of the head. Depending on mood, the stripes may disappear completely, replaced by the dark grey base colour. This is seen mainly in dominant individuals. In sexually active individuals the body becomes almost black, and the yellow gill spot then shows its most intense colour.

Remarks: The Canary Cheek *Tropheus* was first imported only relatively recently, and immediately found fans. The contrast between the dark colour and the bright yellow cheek spot is very striking. As in many other *Tropheus*, the intensity of the coloration diminishes visibly in polluted water. A remarkable phenomenon is seen in sexually active or competing males kept in unlighted tanks with dark substrate and decor - in such situations only the yellow cheek spot is visible. This - or something of the sort - must also be the case in the native biotope, when the fishes spawn in dark corners in the rocky littoral. An observer several metres from the aquarium sees only a dancing yellow spot.

Tropheus sp. Kavala; Population 3.4

Trade name: Tropheus "Kavala", Tropheus "Kipampa".

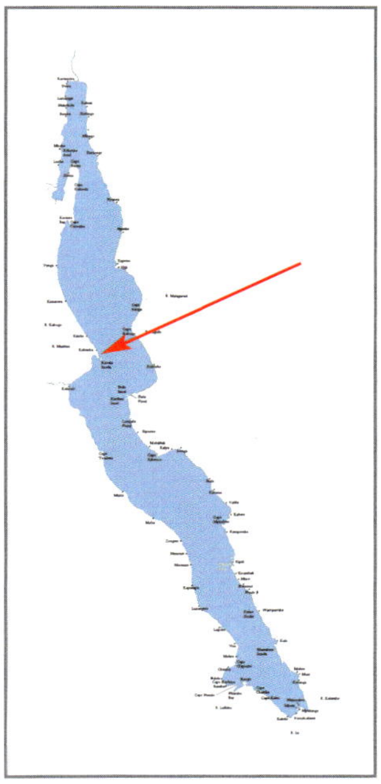

Distribution: The distribution of this colour form begins at Kabimba Bay and extends - including the Kavala islands - to Cape Bwana Denge, a few kilometres north of the Lukuga river, the only outlet of Lake Tanganyika.

Discovered: Like many other *Tropheus* populations in this area, *Tropheus* sp. "Kavala" was discovered during the 1980s by Pierre BRICHARD and his collecting team.

First importation: This local form from the Kavala Islands was collected and imported by Sigmund JANICKI at the beginning of the 1990s.

Coloration: The *Tropheus* from the Kavala Islands has a dark grey base colour with whitish yellow vertical bar the entire length of the body. Apart from the caudal and anal, all the fins are more or less yellow in colour. The first two ray of the ventral and the base of the dorsal are light grey. The anal exhibits a more bluish hue, with numerous yellow eggspots on its posterior part. The upper third of the eye is vivid yellow, and the throat and breast a less brilliant shade of the same colour. The stripes vanish completely in sexually active individuals.

Remarks: The coloration varies somewhat at different collecting sites within the distribution. The most beautiful specimens are supposedly found along a stretch of only 300 metres of coast at Kipampa, which they share with a large population of crocodiles. It would appear that crocodiles are also Tropheus fans, as they are often found in association with the most attractive populations. This local *Tropheus* form presents a splendid contrast between bright yellow fins and dark grey base colour. It very closely resembles the populations on the opposite shore in Tanzania.

Tropheus sp. Chiwa; Population 3.5

Trade name: Tropheus "Chiwa", Tropheus "Kipampa".

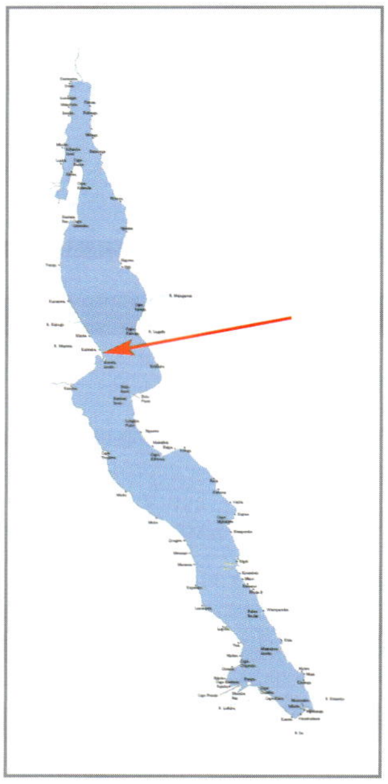

Distribution: The distribution area lies on the mainland to the north of the large Kavala Island, in a small bay south of Kabimba Bay.

Discovered: This population was discovered by Pierre BRICHARD and his collecting team.

First importation: The first importation took place at the beginning of the 1990s, via Sigmund JANICKI.

Coloration: The base coloration consists of a chocolate brown decorated with whitish-yellow vertical bars. Dorsal, pectorals, and ventrals exhibit a vivid yellow with a greenish sheen, which is seen again on the distal part of the caudal. The posterior part of the anal is vivid yellow, while the anterior part is browner. The anterior half of the head is light grey. A narrow brown stripe runs from the eye to the corner of the mouth. The upper third of the eye, and a narrow outer edging, are vivid yellow. The tips of the dorsal lappets are orange, creating the impression of a narrow marginal band. Depending on mood, the light colour of the anterior head may disappear, and then the upper half of the head becomes chocolate brown and the lower assumes a dark bronze sheen, while the yellow coloration of the fins loses some of its intensity.

Remarks: This attractive population was first imported in the early 1990s. This was because, inter alia, these *Tropheus* are found only along a very short stretch of coastline. Brichard certainly knew the spot, but kept quiet about the collecting site, and the situation remained thus for some time until other collecting teams rediscovered the site.
Territorial behaviour is, unfortunately, very pronounced, and if the population is too low problems may occur if these fishes are kept together. The average size of these cichlids is far smaller than we are accustomed to see in other populations. Wild-caught specimens measure 8-9 cm and grow only imperceptibly over a period of a few months. This is thought to be the smallest *Tropheus* known to date.

This attractive *Tropheus* sp. "Chiwa" was first imported at the end of 1995 from the then Zaire. It is found a few kilometres distant from the Kavala population. Females (right) do not have a light forehead. Tropheus sp. "Chiwa" closely resembles the Kigoma population in Tanzania.
photo: P. Schupke

Tropheus sp. Kalemie; Population 3.6

Trade name: Tropheus "Kalemie", Tropheus "Kaniosha", Tropheus "Kioshi"

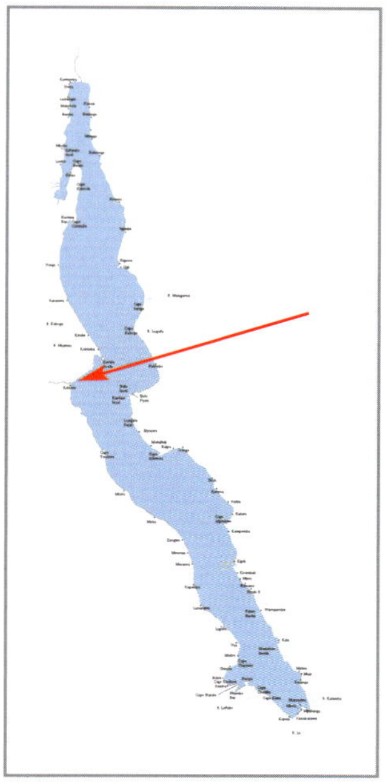

Distribution: The distribution begins at Cape Bwana Denge and continues across the Lukuga past Kalemie.

First importation: The first *Tropheus* of this colour form were imported at the end of the 1980s by Sigmund JANICKI.

Coloration: The body is brown to dark brown-grey with a dirty yellow underside. The head has an indistinct orange sheen. Sub-adult and non-territorial individuals exhibit narrow orange vertical stripes. Depending on locality, the dorsal may be brown or matt orange in colour. The soft dorsal and anal have a metallic blue to blue-green sheen. Most of the operculum is striking bright orange, ditto the upper part of the eye ring. The anal bears a number of eggspots, which may also appear on the ventrals. Juveniles are mainly orange in the dark intervals between the vertical stripes, but unfortunately they lose this attractive coloration with increasing age.

Remarks: *Tropheus* sp. "Kalemie" can have a very high orange component and be very attractive in its coloration. Unfortunately it is very rarely imported. Its large distribution has allowed several small colour variations to develop. In 1900 BOULENGER described *Tropheus annectens* on the basis of specimens that supposedly originated from Kalemie. These *Tropheus annectens* possessed only 4 anal spines, but *Tropheus* sp. "Kalemie" (or *Tropheus* sp. "Kaniosha" as it is also still named) has 6 anal spines. The only *Tropheus* with 4 anal spines caught at Kalemie are similar to *Tropheus polli* to the point of confusion. Unfortunately ignorance of this situation means that *Tropheus* sp. "Kaniosha" or *Tropheus* sp. "Kalemie" is repeatedly referred to *Tropheus annectens*.

Tropheus sp. Tembwe; Population 3.7

Trade name: To the best of my knowledge this local form has no trade names.

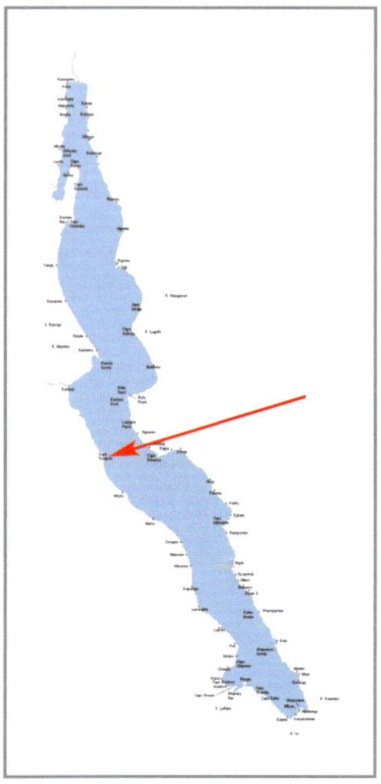

Distribution: The distribution of this colour form is given as Cape Tembwe in the former Zaire. The exact boundaries of its range are unknown.

Discovered: This population of lineage 3 was discovered by Pierre BRICHARD´s collecting team in the mid 1980s.

First importation: As far as I know, this *Tropheus* has not yet been imported for the ornamental fish industry. It is possible that some specimens may have been privately imported by aquarists.

Coloration: The base coloration of this population is - judging from a photograph - dark brown. All the fins are the same brown as the body. Small yellowish spots are visible on the snout and near the eyes. The body probably becomes lighter, depending on mood, as in the population at M¥toto, with the fishes then assuming a yellowish grey colour on the body and the dorsal also paling to the same colour. The upper part of the iris is yellow. Pectorals and ventrals exhibit a rather brownish shade.

Remarks: According to Pierre Brichard, the coast south of Kalemie does not consist of continuous rocky shoreline, but is repeatedly broken by long sandy beaches and river mouths. However, this has apparently not deterred *Tropheus* of lineage 3 from colonising this coastline. The latest research by Christian Sturmbauer and his colleagues indicates that in the past these *Tropheus* had contact with the *Tropheus* at Cape Kungwe. That is hardly surprising, as in those days the south coast of the central lake extended from the Mahale Mountains to the west coast opposite. The research was performed on specimens from Kyeso, which lies a few kilometres south of Cape Tembwe.

High predation pressure, for example from fishing birds like that shown here, can have a dramatic influence on the coloration of individual *Tropheus* populations. **photo:** P. Schupke

Tropheus sp. M´toto; Population 3.8

Trade name: Tropheus "Mtoto", Tropheus "Mikonga", Tropheus "Yellow Nose".

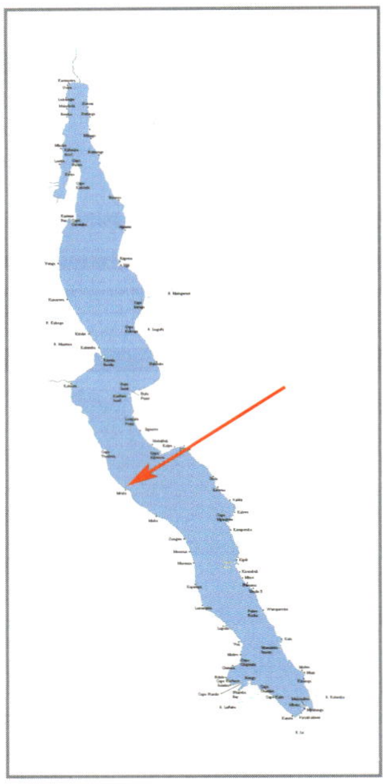

Distribution: This population lives in the vicinity of M´toto and Pala. It is not known how far north its range extends.

Discovered: Pierre Brichard and his collecting team discovered this local Tropheus form at the beginning of the 1980s.

First importation: This *Tropheus* was imported, in not very large numbers, by Sigmund Janicki in the mid 1980s. There have been no further imports due to lack of demand.

Coloration: This population is highly reminiscent of *Tropheus* sp. „Kalemie". Its base colour is brown to dark brown. The few photos in existence show adults without any stripes. Depending on mood and locality, the body may become paler and there are then grey to grey-brown areas on the head and anterior dorsum. There are yellowish areas on the forehead, breast, and belly regions. The pectorals and ventrals are yellowish in colour, while the dorsal and anal are, by contrast, brown-grey or brown. The anal is also very dark, but, as in other members of the lineage, has bluish areas on the soft part, with the odd egg-spot or so. The upper part of the eye is strikingly yellow and the rest of the eye is distinctly silvery. Pierre Brichard states that these *Tropheus* may be more yellowish on the posterior body and the unpaired fins. This coloration depends on locality, and also on mood.

Remarks: *Tropheus* sp. „M´toto", also known as *Tropheus* „Yellow Nose", has the most southerly distribution within lineage 3. According to Brichard, its distribution supposedly lies between M´toto and Pala, where it is found sympatric with a *Tropheus* of lineage 4 that has 4 anal spines and a forked tail. To the north of this area lie more than 100 kilometres of coastline that have not yet been explored. One thing is certain: there must be yet more *Tropheus* of lineage 3 along this section of coast, which probably resemble *Tropheus* sp. „Kalemie" or Tropheus sp. „M´toto".

Tropheus sp. Karilani; Population 3.9

Trade name: Tropheus "Karilani"

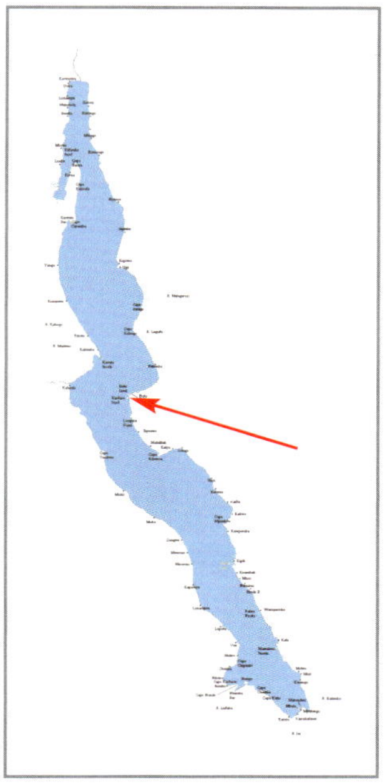

Distribution: The known distribution lies on the north-western side of the Kungwe Mountains in Tanzania, and is said to extend to Halembe, about 40 kilometres to the north-east.

First importation: At the end of the 1980s.

Coloration: The base colour consists of a dark grey-brown. Depending on mood, adult males - unlike juveniles and females - no longer exhibit any barring. The very dark caudal fin contrasts to some extent with the lighter dorsal (whose soft part is yellowish) and anal, which are brown-grey in colour. A narrow orange marginal band can be seen in the dorsal. Depending on mood, these fishes may become light grey on the body, with brownish-pink shades visible on the head. The lips are then dark-bordered. In contrast to the yellowish pectorals, the ventrals are striking yellow. The lower part of the eye ring is silvery, the upper, by contrast, clearly yellow.

Remarks: *Tropheus* sp. "Karilani" exhibits a marked resemblance to *Tropheus* sp. "Kavala" (3.4). They shared a common coastline when Lake Tanganyika was divided into three. In fact they were probably a single ancestral species, rather than two similar spp sharing a common coast. According to data from an importer, *Tropheus* sp. "Karilani" is not found until a few kilometres north of Karilani Island. At the island itself there is, as well as *T. polli, T.* sp. „Kirschfleck", and *T. duboisi*, a *Tropheus* which the collectors call „Karilani Red", and which resembles *Tropheus* sp. „Chipimbi" (7.5) except that it lacks the red band on the caudal peduncle. This local form has not proved very popular with hobbyists on account of its dull coloration, and hence has been imported only a few times. Its aggressive behaviour has likewise done nothing for its popularity. It is quite conceivable that *Tropheus* sp. "Karilani" also occurs south of Bulu Point as the rocky coast extends further south. The southernmost limit of the range of lineage 3 on the Tanzanian shore is at Bulu Point or perhaps somewhat further south.

Tropheus sp. Malagarasi; Population 3.10

Trade name: No trade names known.

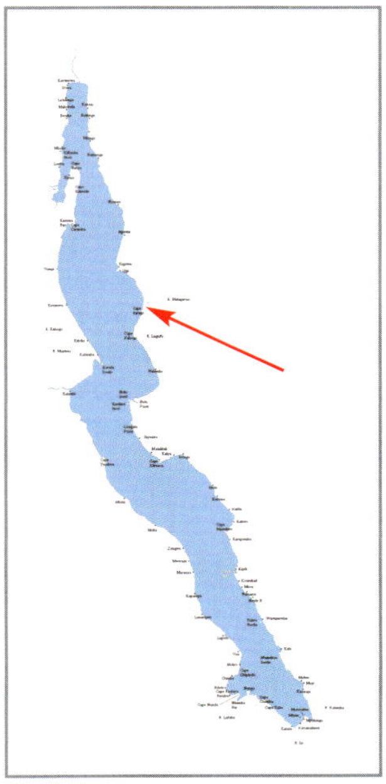

Distribution: The currently known distribution of this local form begins to the north of the Malagarasi Delta in northern Tanzania. It is not known how far north it extends.

Discovered: At the beginning of the 1980s, A. Konings and H. W. Dieckhoff dived along this coast in Tanzania and discovered numerous other local *Tropheus* forms. Ad Konings mentioned this *Tropheus* for the first time in 1988.

First importation: This local form has probably not yet been imported for the trade.

Coloration: The base colour is a dark grey-brown with a slight grey-blue sheen, as can also be seen in neighbouring local forms, eg *Tropheus* sp. "Kavala" (3.4), inter alia. Only dominant adults cease to exhibit bars on the body. The light yellow dorsal has exhibits bluish shades on its posterior soft part, while by contrast the ventrals, as well as the breast and belly, are striking uniform yellow. The pectorals are not as vividly coloured and are rather inconspicuous. The anal has bluish areas on its posterior lobe, and exhibits a number of orange egg-spots. The eye ring is silvery, but with yellow to orange hues in its upper part.

Remarks: Little is known about this local form, because it probably has not yet been imported. It is not known whether it occurs with other *Tropheus* lineages or species. *Tropheus* of lineage 5 and *Tropheus duboisi* are found only a few kilometres further south. *Tropheus* sp. "Malagarasi" exhibits a close similarity to *Tropheus* of this lineage occurring further to the north. Males become much darker on the body when sexually active, and no longer have stripes during this phase. The stripes may disappear for long periods of time, and only if the fish´s situation changes in the long term does a slight barring reappear.

Tropheus sp. Ujiji (Katonga); Populations 3.11 and 3.12

Trade name: Tropheus "Ujiji", Tropheus "Katonga".

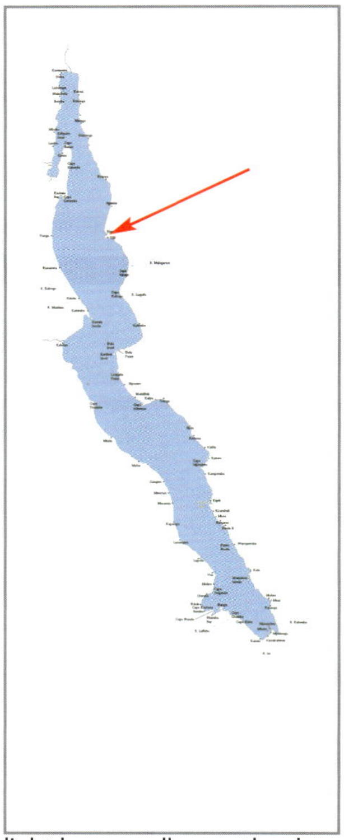

Distribution: The distribution begins a few kilometres south of Kigoma and ends a few kilometres south of Ujiji, where a swampy area has formed around the mouths of several small rivers.

First importation: These extraordinarily attractive local forms were already being imported, albeit at long intervals, at the end of the 1980s.

3.11. (Ujiji)

3.12. (Katonga)

Coloration: Dominant males have a dark grey-brown base colour with residual traces of the whitish-yellow vertical barring. In sexually active individuals the barring may disappear almost completely. A light lemon yellow area extends some distance down the side from the base of the rather yellowish dorsal, which has an orange marginal band. The ventrals and anal, as well as the breast and belly, are the same light lemon yellow as the dorsal. The soft part of the anal has an attractive bluish turquoise zone set with numerous orange egg-spots. The tips of the spines are likewise orange. The silvery eye ring has a faint light blue sheen and is yellowish in its upper part. Juveniles and females exhibit whitish to whitish-yellow vertical bars, as in all *Tropheus* of this lineage.

Remarks: These two populations are very similar to one another and appear to be linked by transitional forms, which is why they (and the following two populations) are dealt with together here. The most attractive individuals (Population 3.12) are supposedly caught at Katonga. In the southern part of the range of this colour form, towards Ujiji (3.11) the lovely dorsal patch and the light yellow zone on the belly are greatly reduced, with the bright yellow remaining only on the breast. Unfortunately these attractive *Tropheus* have proved to be very belligerent, and this can sometimes lead to almost insoluble problems when maintaining them in the aquarium. But this will be no obstacle, but rather a challenge, to confirmed *Tropheus* fans and those who aspire to that status. *Tropheus duboisi* also lives along this stretch of coast, but in rather deeper water. It is not known whether other *Tropheus* lineages also occur here. In all probability there may be *Tropheus* of lineage 5 (blue-eyed greens) in this region.

Tropheus sp. Kigoma I (Nyanza Lac II); Populations 3.13 and 3.14

Trade name: Tropheus "Kigoma", Banded Tropheus.

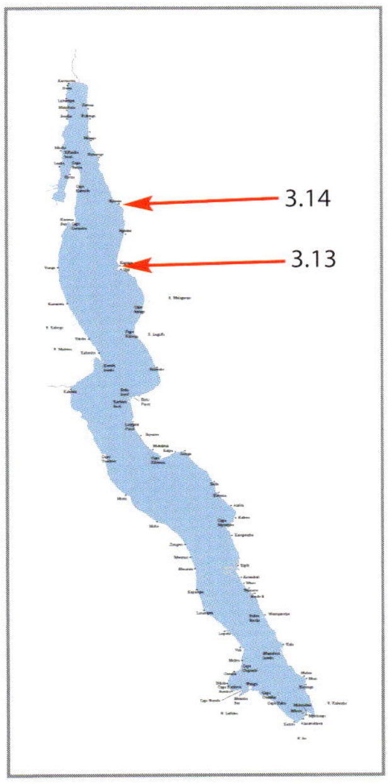

Distribution: The distribution of this local form begins somewhat to the south of Kigoma and ends at Nyanza Lac in the extreme south of Burundi.

Discovered: The *Tropheus* at Kigoma were discovered by W. Staeck in 1974 and first documented in the literature in the same year.

First importation: First imported for the ornamental fish industry in the autumn of 1975.

3.13. (Kigoma I)

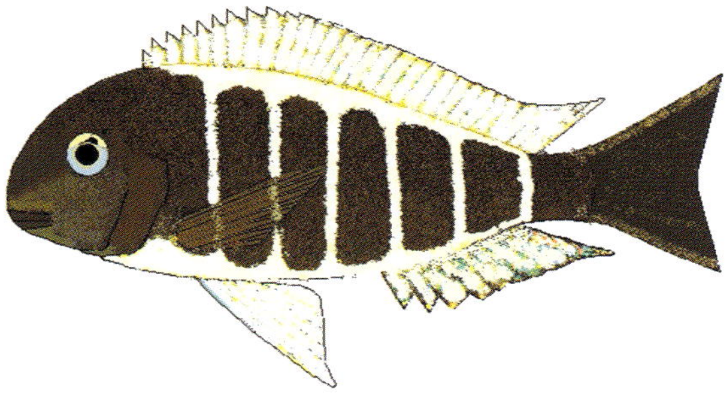

3.14. (Nyanza Lac II)

Coloration: The base colour consists of a brown-grey. The ventrals are yellow and the pectorals almost colourless. Dorsal and anal are more grey, with a faint yellow sheen. The anal, which tends more to grey-yellow, has a slight bluish cast and has a few orange-yellowish egg-spots on its soft part. Unlike the grey-brown sexually inactive individuals, sexually active males are very strikingly coloured: their white dorsal, anal, and ventral fins are clearly visible from a considerable distance. Juveniles and females exhibit up to eight vertical bars, which are often also still to be seen in males. The eye ring is light silver with a faint bluish tinge, and its upper part is yellowish. The sub-population at Nyanza Lac again shows rather more colour: the dorsal, ventrals, belly, throat, and anal are again rather yellower than in individuals from the vicinity of Kigoma.

Remarks: At Nyanza Lac as at Kigoma, there are *Tropheus* of lineage 5 (blue-eyed greens) as well as those of lineage 3. In addition, Nyanza Lac is also home to the *Tropheus* „Chocolate" or saddle-spot *Tropheus*, which belongs to lineage 2 and has been scientifically described as *Tropheus brichardi*. These three lineages are easy to differentiate in the aquarium (see key to the lineages), although Brichard regarded them as one species and spoke of the extraordinary variability of the *Tropheus* at Nyanza Lac. In the lake, however, there can be problems in distinguishing these *Tropheus* from one another or even detecting that there are different species involved. Their neutral coloration can be very similar and it is often possible to tell whether the stripes are greenish-brownish or brownish-yellowish only when they are viewed in the aquarium.

Tropheus sp. Bulombora, Population 3.15

Trade name: Tropheus "Bulombora".

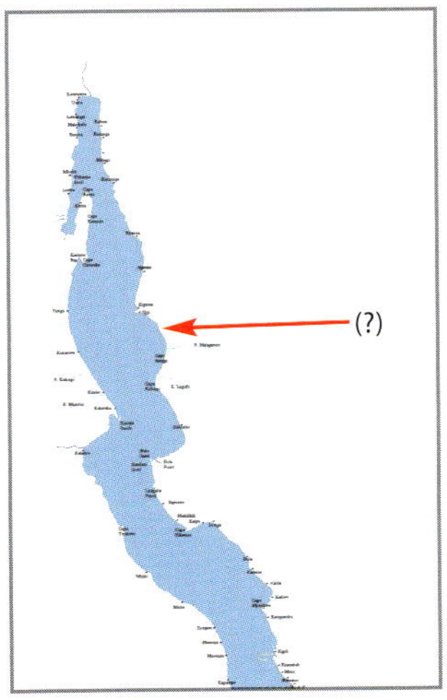

Distribution: The collecting locality is given as Bulombora by the importers, but I know of no village of this name. There is however, a village with the very similar-sounding name of Ulombola approximately in the middle of the stretch of coast between the Malagarasi delta and Kigoma.

Discovered: This colour form was first discovered in 2001, purportedly by one of the local collecting companies.

First importation: As far as I know, the first importation took place in autumn 2001.

Coloration: The base colour of the body in sexually mature individuals consists of a brownish yellow-green. The underside is yellow. Males have a yellow-white saddle spot on the anterior half of the dorsum. Apart from the dark caudal, all the fins are bright yellow in females and juveniles and overlain with brown in males, apart from a yellow-white spot in the dorsal above the saddle. Dominant individuals of both sexes have yellow to yellow-white vertical stripes on the body. Usually only hints of these bars are visible in females, whose body is a yellowish green colour. The light-coloured iris has a yellowish hue in its upper third, which may be bisected by a dark bar. The upper and lower lip area is blackish. The forehead may be green-turquoise in males. The ochre yellow fry resemble those of lineage 9, at least up to a size of about 3 cm.

Remarks: This local variant is one of the few *Tropheus* species and their colour forms in which the sexes can be differentiated on the basis of coloration. Males always exhibit the saddle-spot, while the generally yellow-green females, by contrast, do not have this marking. This attractively coloured *Tropheus* was first exported for the trade in autumn 2001.

Pair spawning; male to the left.

Pair spawning; female to the left. **photo**: P. Schupke

Tropheus of lineage 4

This *Tropheus* lineage is found along the central east and west coasts. Its current distribution in the Congo supposedly extends from the Kavala Islands, about 30 kilometres north-west of Kalemie, down as far as Moba. In Tanzania its distribution begins about 20-30 kilometres north of Ikola and extends at least as far as Karilani Island, which lies to the north-west of the Mahale Mountains.

In its natural habitat it comes into contact with various other lineages. In the north-west part of their range, at the Kavala Islands, they share the coastline with Tropheus of lineages 3 and 5. They have been shown to occur at Kyeso, south of Cape Tembwe. At Moba in the south they come into contact with the northern „Rainbow" *Tropheus*. The first Tropheus (Pop. 4.1) of this lineage were discovered on the Tanzanian shore and subsequently described as *Tropheus polli* by Axelrod, 1977. They live along a long stretch of coast which encompasses the Mahale Mountains and extends into the large bay at whose centre lies the village of Ikola. Their occurrence in the northern part of this bay, where they are found sympatric with *Tropheus* sp. „Ikola" (lineage 12), has only recently been discovered. In the northern part of this large bay *T. polli* occupies the upper rocky littoral while *T.* sp. „Ikola" has been forced down to the deeper layers of water. From Kalia, which lies at the end of the bay, to Cape Kibwesa they live sympatric with a *Tropheus* population (6.3) that resembles the *Tropheus* from Cape Kibwesa. At the northern part of the Kungwe Mountains they share a section of coastline with *Tropheus* of lineage 10, *Tropheus* sp. "Karilani" (3.9), and *Tropheus* sp. „Kirschfleck" (lineage 12), but occupying separate habitats. On the west coast, in the Congo, the distribution of the other two populations (Pops. 4. 3 and 4. 2) have not been fully documented. The appearance of these *Tropheus* resembles that of the east coast form that has been described *Tropheus polli*, except that they are somewhat darker, more brown-black, while *Tropheus polli* is overall lighter and appears more grey. Juveniles of populations 4.1 and 4.3 resemble each other and can be confused. The females too have pronounced vertical bars which they generally retain. Dominant and sexually active individuals become dark and lose their stripes. If an individual loses its territory, then the stripes reappear.

The *Tropheus* of this lineage are notable for having a caudal fin that is deeply forked in a sickle shape and has prolonged pointed tips. In individuals from Moba (4.2) the caudal is likewise sickle-shaped, but the tips are not as prolonged (in individuals that are not yet full grown). But the overall appearance is very similar to those further north. All share the common feature of four spines in the anal fin. *Tropheus annectens*, described by Boulenger in 1900, likewise possesses only four anal spines and was supposedly collected in the vicinity of Kalemie. For a long time the only *Tropheus* known from the Kalemie area was a brown population of lineage 3, which has been incorrectly identified as *Tropheus annectens* by a number of authors. It was not until much later that further *Tropheus*, with four anal spines, were discovered between Kalemie und Moba. The *Tropheus* of lineage 4 attain a larger average size than those of other *Tropheus* lineages. Including the much elongated tail, males of populations 4.1 and 4.3 can attain the 20 cm mark. At present I have no details of the ultimate size of individuals of population 4.2 at Moba.

Tropheus polli was collected for the ornamental fish industry at the beginning of the 1980s. The other two *Tropheus* of this lineage were not discovered until later and have been imported in smaller quantities.

This lineage contains the populations numbered 4.1 - 4.3.

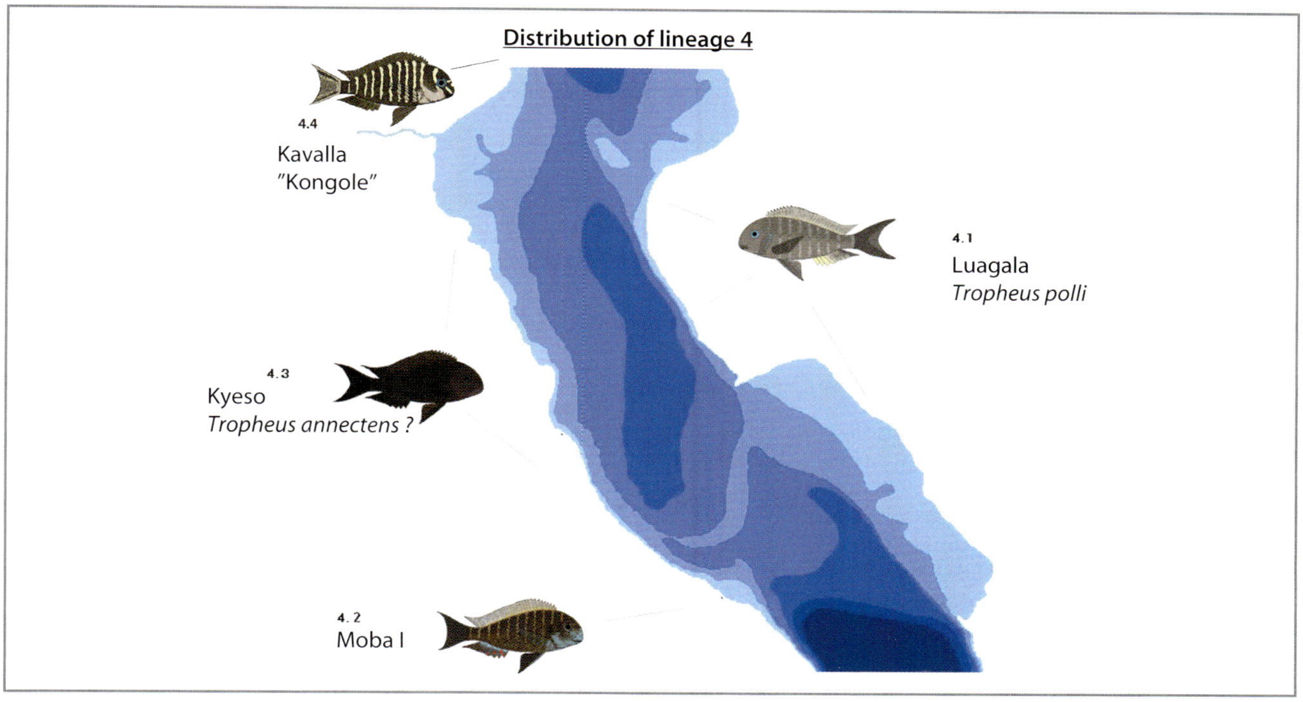

Tropheus polli Axelrod, 1977 Population 4.1

Trade namen: Tropheus polli, Fork-tailed Tropheus

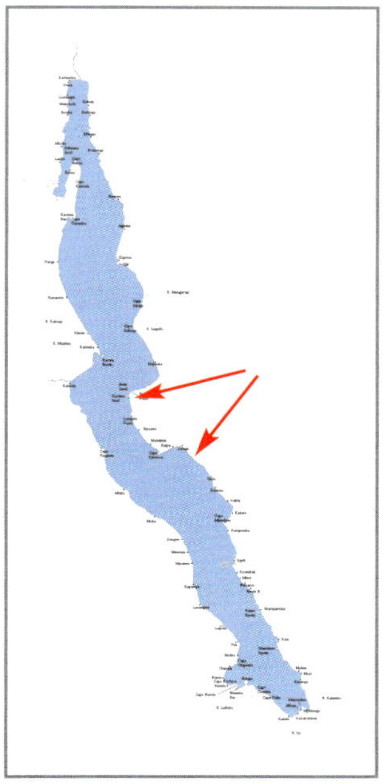

Distribution: The currently known distribution in Tanzania extends from the northern side of the Mahale Mountains, where the Bulu Islands are situated, past Luagala Point and into the northern part of the range of *Tropheus* sp. „Ikola" (11.1).

Discovered: This *Tropheus* was discovered in 1975 by Fainzilber and described by Herbert Axelrod as *Tropheus polli* in 1977.

First importation: *Tropheus polli* was first imported for the trade at the end of the 1970s/beginning of the 1980s. It has been available from time to time, albeit at long intervals, up to the present day.

Coloration: The base coloration consists of a light grey, which can become darker depending on mood. Breast, belly, ventrals, anal, and the tips of the caudal may have a more or less yellow cast. In some individuals the anal may be a brighter yellow. Only dominant individuals lose their stripes virtually completely. In general, however, the stripes are less pronounced in males than in females and juveniles, whose banding is very contrast-rich. Dorsal, pectorals, and caudal are usually the same colour as the body. The caudal and anal may sometimes have a few faint red streaks and dots. The iris is completely light blue apart from a darker spot in its upper part.

Remarks: *Tropheus polli* has been known for a long time in the aquarium hobby and is regarded as rather more peaceful than other members of the genus. The large forked tail is a notable feature of these fishes. A relative from the same lineage occurs to the north-west of Luagala Point in the Congo and is dark brown rather than grey in colour. The two form can easily be confuses as juveniles, as the barring and contrast are practically identical. These *Tropheus* have been sold as *Tropheus* sp. "Kongole" (see 4.3). In the past, when Lake Tanganyika was divided into 3 lakes, these two forms shared a common coastline in the central lake, which explains the marked similarity between them. *Tropheus polli* prefers to live at a depth of five to ten metres, which is significantly deeper than the preferred depth for other *Tropheus*, with the exception of *Tropheus duboisi*. To the north of Ikola *T. polli* has forced *T.* sp. „Ikola" (11.1) into deeper water. At

Kalya, between Isonga and Cape Kibwesa, it lives with *Tropheus* sp. "Kalya" (Pop. 6.4) in an underwater ravine immediately beneath the water´s surface. It is not known whether or to what extent *Tropheus polli* has formed local colour forms within its rather large distribution area. Any differences cannot be great, as otherwise they would probably have been documented - the area in which the species occurs has been relatively well studied.

Tropheus sp. Moba I; Population 4.2

Trade name : Tropheus "Moba".

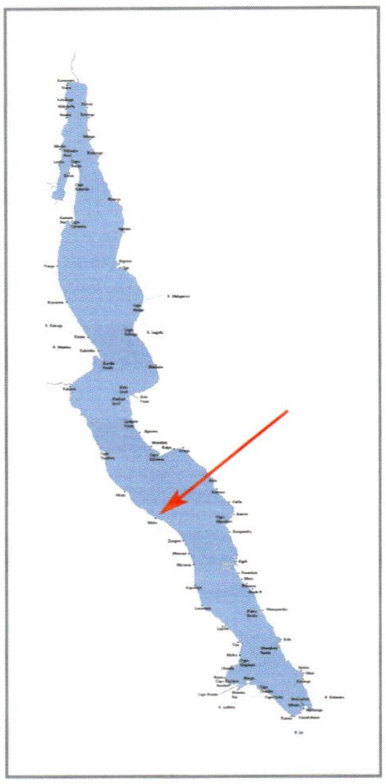

Distribution: The currently known distribution extends from Moba Bay northwards to Pala.

Discovered: This *Tropheus* was probably discovered by Pierre Brichard´s collecting team at the beginning of the 1980s.

First importation: The first individuals (pond-bred) were exported for the trade by Pierre Brichard at the beginning of the 1990s.

Coloration: The base coloration ranges from brownish to brownish-green. Individuals in neutral mood exhibit narrow, whitish to ochre-coloured, vertical bars, which may disappear when the fish becomes excited. The snout, mouth, cheek, throat, and breast areas, as well as the base the pectoral fins, are conspicuously turquoise. The dorsal is faint yellowish-grey and thus rather lighter than the rest of the body. There are small pink spots on the cheek area. The anal, particularly the spinous part, has small orange spots and streaks. Caudal, ventrals, and pectorals are inconspicuously coloured. The iris is blackish, with no additional colour. I have no information on the colour, and particularly the markings, of juveniles.

Remarks: *Tropheus* sp. "Moba" lives sympatric with *Tropheus* sp. "M´toto" (lineage 3) between M´toto and Pala. It can be assumed that these two *Tropheus* species live at different depths in the local rocky habitat. It is not beyond the bounds of possibility that *Tropheus* sp. "Moba" encounters northerly "Rainbow" *Tropheus* of lineage 8 in the south of Moba Bay. Unfortunately Brichard provides no data regarding this possibility. Unfortunately the pond-bred specimens I studied grew hardly at all, attaining a length of only about 10 cm, so that no assumptions could be made regarding ultimate size. This *Tropheus* may be a colour form of *Tropheus* population 4.3.

Hippos are the most dangerous (for humans) of the large animals of Africa, and it is impossible to collect aquarium fishes where they occur. This photo was taken at Cape Nundo. **photo:** P. Schupke

Tropheus sp. Kyeso; Population 4.3

Trade name: No trade name exists for this local form.

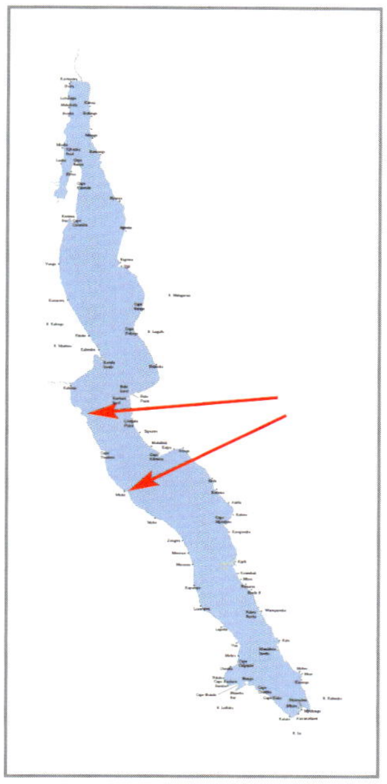

Distribution: The currently known distribution of this *Tropheus* lies between the village of Kyeso, south of Cape Tembwe, and a point a few kilometres south of M´toto.

Discovered: It cannot be stated for sure if this form is the species described by BOULENGER as *Tropheus annectens*, as the origin of the *Tropheus* that BOULENGER received from Captain HECQUE is not specified (probably from a fish marked at Kalemie).

First importation: The first specimens of this *Tropheus* to be imported were pond-bred and originated from Pierre BRICHARD. Wild-caught specimens followed a short time afterwards, imported by CJ Aquarium.

Coloration: The base coloration is dark brown in large specimens, with the edge of each scale lighter. Light vertical bars are present only on the back below the dorsal. The head is lighter and has a slight pink cast. The mouth, anal, and the posterior edge of the gill-cover as well as the caudal membranes and the base of the dorsal, exhibit shades of grey-blue. The iris is dark with a slight bluish sheen. Juveniles have a contrasting colour pattern of dark brown and whitish beige, and the iris is mainly light blue with a small red area on the upper edge of the eye.

Remarks: Unfortunately not much is known about this *Tropheus*. Because its distribution is immediately adjacent to the former northern lake, it would be interesting to know whether this Tropheus has been able to spread north - and how far. The same applies to the southern part of its range, the other side of Kalemie. This area of coastline includes numerous sandy beaches and only small, discontinuous, rocky shores. Still further to the south rocky shoreline becomes more common, and this is the home of population 4.2, which is very similar in appearance to *T.* "Kongole". The coast between Kalemie and Moba undoubtedly harbours many such puzzles. The political situation in the Congo makes it far from easy to travel there in safety and to study the terrain in the detail it deserves. The coastline of the former Zaire is almost as long as Lake Tanganyika itself. According to the latest DNA research this population is one of the oldest *Tropheus* lineages.
After the editorial deadline it came to the author´s knowledge that another population of lineage 4 was collected in the 1980s by Pierre BRICHARD near the Kavalla islands in Congo. It is catalogued here as population **4.4 Kavalla "Kongole"**.

Tropheus of lineage 5

This lineage occupies a very long section of coastline, extending from the extreme north, at Kavimvira in the former Zaire, to the Kavala Islands. On the Tanzanian shore the distribution begins in the north at Nyanza Lac, perhaps even at Rumonge in southern Burundi, and extends - as far as we know at present - at least as far as Halembe. With a few exceptions, almost all of the locations for the "Blue-Eyed Greens", as these *Tropheus* are sometimes known, are also home to representatives of other *Tropheus* lineages (species). At Kavimvira in the extreme north-west they encounter *Tropheus* of lineage 2. On the southern half of the Ubwari peninsula and at Rumonge they are sympatric with lineage 1. From Kasimia Bay to Kunanwa they are alone, and according to BRICHARD their population density is very high in this area. From Kunanwa they occur with lineage 3 south to the Kavala Islands, supposedly the southernmost extreme of their range. From Nyanza Lac to Kigoma in Tanzania this lineage is found with lineage 3 and lineage 10 (to which *Tropheus duboisi* belongs). At Nyanza Lac there is also a fourth lineage, lineage 2, which includes *Tropheus brichardi*, also known as the Chocolate Moorii. According to KONINGS (1988; p.124, third photo from top left) the southernmost locality is at Isonga, but scientists from the University of Innsbruck were unable to find the *Tropheus* in question at Isonga.
The reason for assigning these *Tropheus* to a lineage of their own lies in their external appearance. The base colour consists of various shades of green, which, depending on mood, may sometimes verge on brown-green. The second important colour associated with green is yellow, and this colour is expressed in the stripes, as spots, and as narrow bands. The eye is always blue in colour, in all populations. A very important feature separating these *Tropheus* from others is the differently shaped caudal, which is large, deeply forked, and with slightly prolonged tips. Because of damage during shipping, this character is often no longer visible in newly imported individuals. It appears where the Blue-Eyed Greens are found with lineage 3 their numbers are subordinate to the other *Tropheus*. BRICHARD mentions this several times in his 1989 book "Cichlids and all other Fishes of Lake Tanganyika". Between Kasimia Bay and Kunanwa they live alone, ie with no other *Tropheus*, and here their population density is supposedly higher. Moreover, no migrations

due to fluctuating water levels took place here. This may lead us to conclude that lineage 5 is an older lineage that is being slowly overwhelmed by the younger lineage 3. Recent research by the University of Innsbruck has shown that the „green" *Tropheus*, at least those found between the Malagarasi and Cape Kabogo, are the second oldest lineage. This may signify that long ago these *Tropheus* and *Tropheus duboisi* were the only *Tropheus* inhabiting this part of the lake. The lineages 1, 2, and 3, which today occupy this region, are younger. "Green" *Tropheus* from the west coast could not be studied. According to our present knowledge, lineage 5 may have evolved in the northernmost of the three former lakes. Unfortunately the central western distribution between Kasimia Bay and Kabimba is not a popular destination for collecting teams, who where possible avoid this stretch of coast because, as in many other parts of the former Zaire, the military are often very arbitrary in their actions, and freedom can quite unexpectedly hang by a thread, as the author himself found out in Kananga, in the central Congo. Moreover, piracy is on the increase on the lake, and the west coast is a favourite haunt of pirates.
This lineage contains the populations numbered 5.1 - 5.14.

Distribution of lineage 5

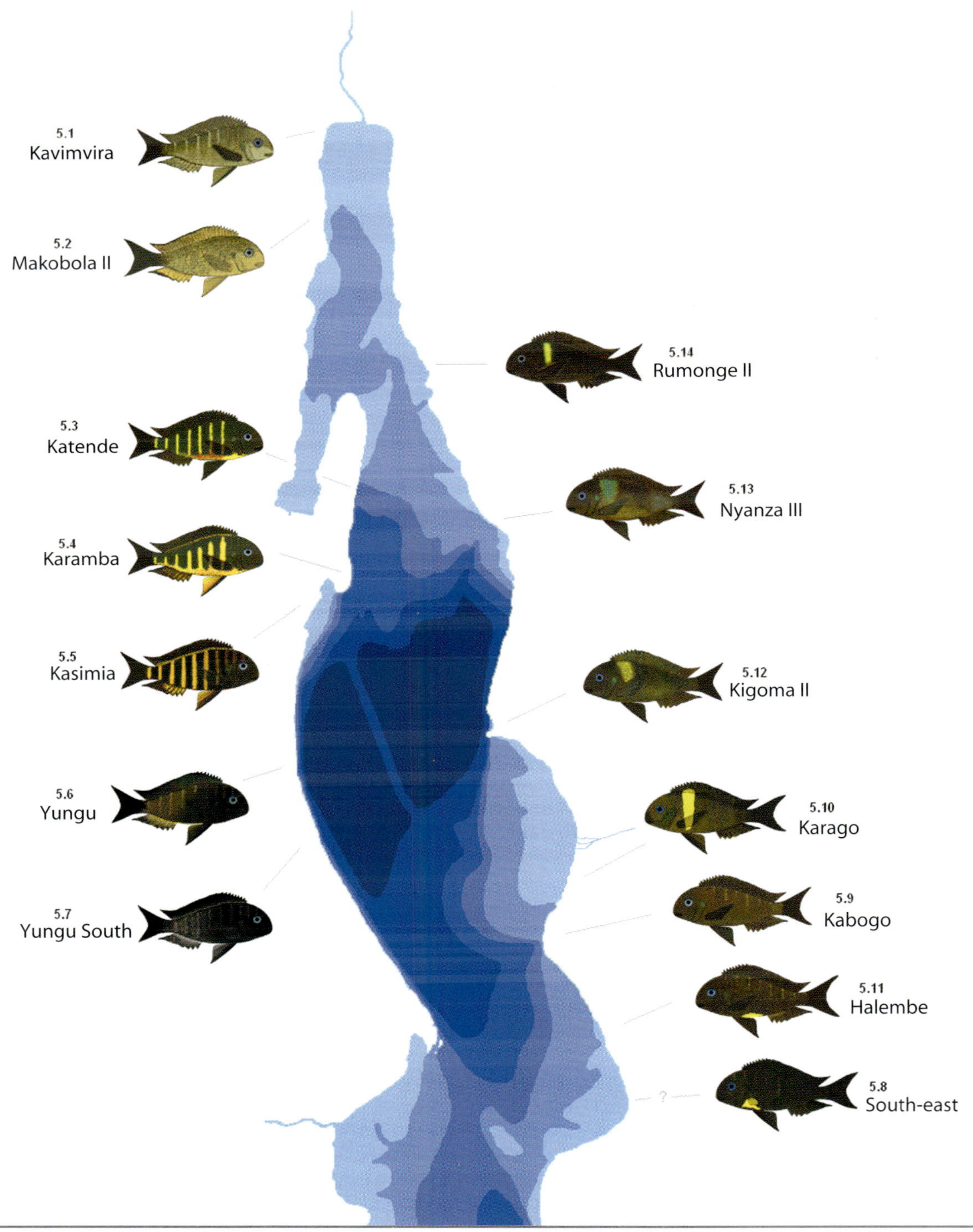

Tropheus sp. Kavimvira; Population 5.1

Trade name : Tropheus "Kavimvira green".

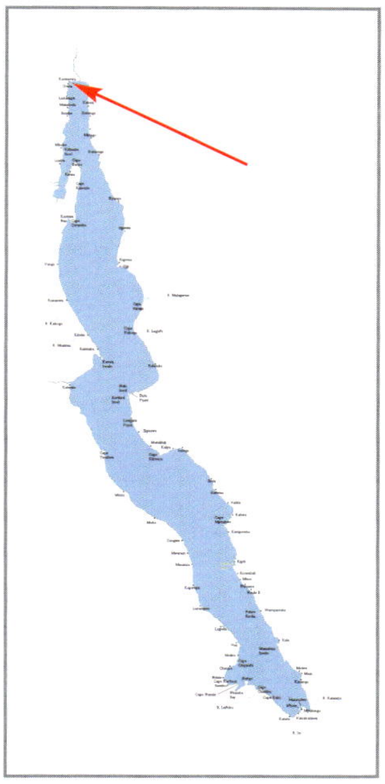

Distribution: : This *Tropheus* lives in the extreme north-west of the lake at Kavimvira in the Congo.

Discovered: This local form was discovered by Pierre BRICHARD´s collecting team at the beginning of 1984.

First importation: It is not possible to say for certain when the first importation took place, but it was probably in the mid 1980s.

Coloration: The base colour consists of a light green-beige-yellow. The lower half of the body, together with the operculum and cheek, is a lot lighter - the colour can best be described as a dull beige-yellow. In this local form vertical bars are clearly visible only on the posterior half of the body; they are light yellow and frequently fragmented. Apart from the straw-coloured anal and the yellowish ventrals, the fins are the same colour as the base colour of the body. The light blue iris is characteristic of lineage 5, and seen in all populations. Depending on mood, the body may become much darker in dominant individuals, which then appear dark green with a lighter belly region overlain with a yellowish hue. The yellow stripes are retained. In undamaged adults the caudal is deeply forked, a common feature of all *Tropheus* of this lineage. Juveniles are beige-yellowish to yellowish-green and have yellowish stripes all over the body.

Remarks: *Tropheus* sp. „Kavimvira" lives sympatric with *Tropheus* of lineage 2 (Pop. 2.3). It is very much subordinate to lineage 2 in terms of numbers, and hence it may be assumed that it is being supplanted by the younger lineage 2. The sediment of the Ruzizi river and (or) the fluctuating water level of Lake Tanganyika in the past has brought these two lineages together on this part of the north-west coast. The *Tropheus* of lineage 5 appear to be the losers. In the mid 1990s this local form appeared on stock lists in the ornamental fish industry in Germany for a short time, but the demand from aquarists was limited and the form was thus correspondingly rarely imported.

Tropheus sp. Makobola II; Population 5.2

Trade name : Tropheus "Makobola".

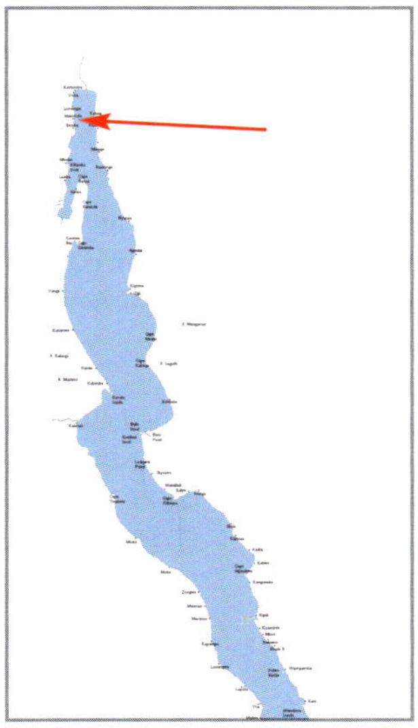

Distribution: This local *Tropheus* form is found in the north-west of the lake at Makobola, and its distribution extends further to the north.

Discovered: Pierre BRICHARD´s collecting team discovered this *Tropheus* colour variant at the beginning of the 1980s.

First importation: The first importation took place at the beginning of the 1990s.

Coloration: The base colour consists of a yellowish-green-beige shade, somewhat reminiscent of bronze. Cheek, gill-cover, belly, and anal are yellower in colour. The dorsal is the colour of the body. The light edging to each scale creates a slight reticulated pattern on the body. The iris is a light blue colour.

Remarks: In terms of external appearance, *Tropheus* sp. "Makobola" has many features in common with *Tropheus* sp. "Kavimvira" (5.1), also in this lineage. The few, not particularly good, photos in existence show both species with the large, deeply forked caudal so characteristic of lineage 5. In the region north of Makobola *Tropheus* sp. "Makobola" is sympatric with *Tropheus* of lineage 2 (more precisely, population 2.5), which has already been discussed earlier. The *Tropheus* of lineage 2 are highly variable in their coloration so that the coloration of lineage 5 is not particularly conspicuous - BRICHARD speaks of a melting-pot of variants in this area. However, the differently shaped tail should have given pause for thought. The *Tropheus* to the north of Makobola are not a single species or hybrids of a number of variants. If different *Tropheus* populations were able to hybridise so easily, then, because the same conditions have obtained in the lake for several thousand years (albeit a short period, geologically speaking), all the different forms that once existed in any spot would by now look exactly the same. But that is clearly not the case. According to BRICHARD there is still a link between Makobola and Bemba. That is interesting, as in that case it would theoretically be possible for *Tropheus* of lineage 5 to still live there, even though they have not yet been found. So far only *Tropheus duboisi* and *T.* sp. „Bemba" (Pop. 2.4) have been found at Bemba.

Tropheus sp. Ubwari; Populations 5.3 and 5.4

Trade name : Tropheus "Ubwari green".

5.3 Katende

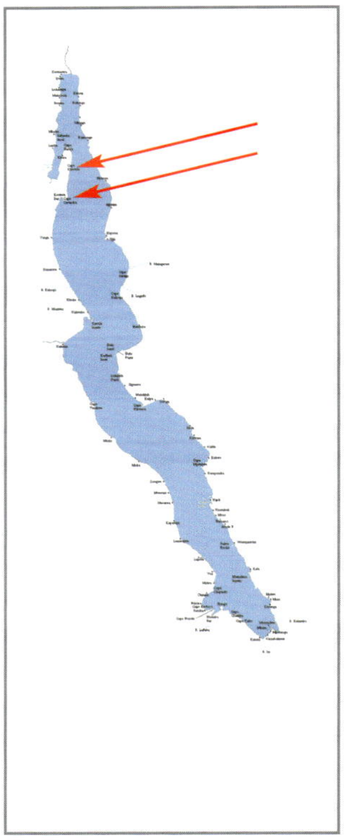

Distribution: The distribution extends from Cape Banza to Cape Karamba, encompassing a stretch of rocky coast about 60 kilometres long.

Discovered: *Tropheus* sp. "Ubwari" was discovered by Pierre BRICHARD´s collecting team back in 1981.

First importation: I do not know if or when wild-caught specimens were first imported. The first *Tropheus* sp. "Ubwari" I saw, at the beginning of the 1990s, were pond-bred specimens at Marc DANHIEUX´s company MalTaVi.

5.4 Karamba

Coloration: The base colour consists of a greenish brown, which can become much darker depending on mood, and then looks rather more like dark brown. Individuals from the vicinity of Cape Katende are greener. The yellow vertical stripes are seen only on the posterior half of the body in adult males, and are only faintly discernible on the anterior part. This may be seen in females as well, but to a lesser extent. Depending on the locality, the yellow stripes may contain orange-red scales or even be completely orange. There are, however, situations - eg during sexual activity - where the stripes may disappear completely. Apart from the dorsal and anal, which are yellow at the base, the fins are all dark in colour. The iris is blue. The caudal is - obligatorily for lineage 5 - deeply forked in adults, and its tips - if intact - are somewhat prolonged. Juveniles exhibit narrow yellowish vertical stripes.

Remarks: *Tropheus* sp. "Ubwari" can be very attractive in appearance and compares favourably with other *Tropheus*. Pierre BRICHARD realised this at an early stage: he collected adult specimens and bred them in large open-air ponds. It was not possible to obtain wild-caught specimens from him. There are no pictures of this attractive fish in his books - he was always at pains to avoid providing his competitors with information. This, of course, was to the detriment of the aquarist, who had to make do with pond-bred fishes, which were often problematical to rear. Moreover, I am of the view that racial purity was not always guaranteed. Another interesting point, in my opinion, is the fact that unfortunately Ubwaris appear to be rather susceptible to the sometimes very persistent "Tropheus disease" that is always lurking in the wings. For some incomprehensible reason these attractive *Tropheus* are rarely imported. In my view they seem no more aggressive than other *Tropheus*.

The green sheen, barring, and colour at the base of the dorsal and anal are indicative of *Tropheus* sp. "Ubwari", while the high red component, particularly on the gill-cover points to *Tropheus* of lineage 1. This fish originated from pond culture in Burundi. Presumably it is a hybrid betwen *Tropheus* sp. "Ubwari" and *Tropheus* sp. "Karamba" (Pop. 1.9). **photo**: P. Schupke

Tropheus sp. Kasimia; Population 5.5

Trade name: I know of no trade name for this Tropheus.

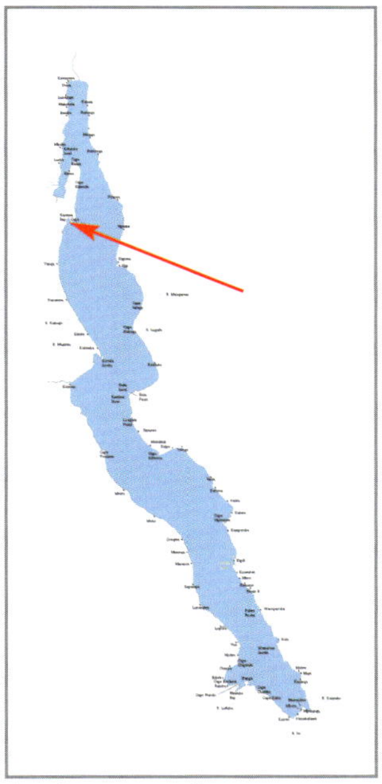

Distribution: According to the importer. these *Tropheus* were captured in Kasimia Bay, which separates the southern end of the Ubwari peninsula from the mainland.

Discovered: This *Tropheus* was probably discovered by Pierre Brichard, as it appeared in the trade as pond-bred.

First importation: I found these fishes at an importer´s at the beginning of the 1990s. I cannot say whether this local form was subsequently imported wild-caught, under this or a different name.

Coloration: *Tropheus* sp. "Kasimia" is very similar to the Ubwari local form (5.3 and 5.4). Its body pattern and coloration are distributed in much the same way. The base colour again consists of a greenish-brownish shade, which, depending on mood, can become rather light, and during a number of activities, eg courtship, turn much darker. The intensity of the coloration decreases considerably towards the end of the body, unlike the yellow stripes, which become more prominent from the middle of the body backwards. This peculiarity is also seen in females Ubwari *Tropheus*. The light blue iris and forked tail typical of all members of lineage 5 are seen in this form as well.

Remarks: The importer informed me that this local form is found in the immediate vicinity of Kasimia Bay. Its distribution area is thus not very large. In this case too the fishes I maintained originated from Brichard´s breeding complex. As with other *Tropheus* from this source, their subsequent growth was insignificant, and they attained an eventual size of barely 8 cm. Their readiness to breed was only minimal, and they produced only 3 fry in the space of a year. The probability of these *Tropheus* being imported again is slight, as the political situation in the former Zaire remains chaotic and any collecting trip would be incalculably risky.

Tropheus sp. Yungu; Populations 5.6 and 5.7

Trade name: Tropheus "Yungu" (5.6), Charcoal Moorii (5.7)

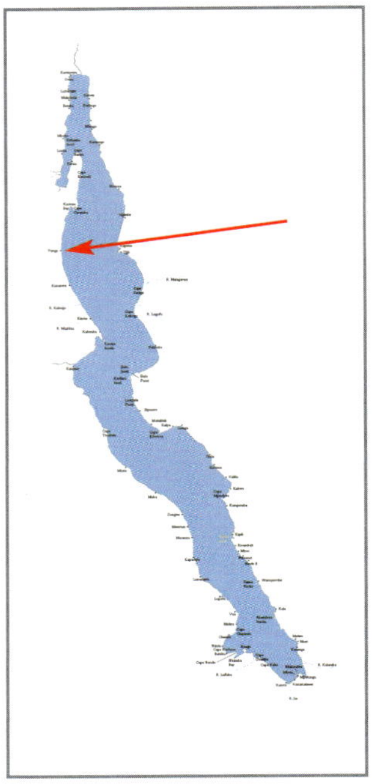

Distribution: The distribution of *Tropheus* sp. "Yungu" lies close to the village of the same name. The site for the Charcoal Moorii is unknown.

Discovered : This local form (Pop. 5.6) was discovered in 1984 by Pierre BRICHARD and his collecting team.

First importation: The first importation of *T.* sp. "Yungu" was by the company C. J. Aquarium and Helmut LÖFFLAD, in 1995. The Charcoal Moorii was imported as long ago as 1986.

Coloration: The normal base colour consists of a greenish brownish shade, broken by up to nine yellowish vertical bars, which are more prominent on the upper half of the body in adult specimens. In juveniles these bands extend onto the lower body as well. Depending on mood, the brownish colour may vanish leaving a greenish to jade-colour shade. The dorsal is verdigris- or copper-coloured. Apart from the almost black caudal and the bright yellow anal, the fins are inconspicuous. Sexually active males become much darker on the head region. The dividing line in the coloration is not vertical, but begins round about the first dorsal spine and runs diagonally to just in front of the first anal spine. In this colour phase the light blue iris veritably glows. The posterior part of the body remains lighter. The deeply-forked caudal is striking. Population 5.7 has no yellow component. Sexually active males are black rather than dark brown, paling to dark anthracite grey further back. The iris is light blue in this form as well.

Remarks: Between Kasimia Bay and the Kavala Islands, this population at Yungu is the only one on which there are a few data. As regards the rest of the coastline, it is known only that *Tropheus* of lineage 5 occur alone from Kasimia Bay to Kunanwa and attain a rather high population density. From Kunanwa to the Kavala Islands and a few kilometres further south they encounter *Tropheus* of lineage 3, and here there population density is supposedly lower. It may be assumed that the coloration of these *Tropheus* must vary at least to some degree along this approximately 80 kilometre stretch of coast. This can be clearly seen from population 5.7, which was imported in 1986 but found few fans because of its inconspicuous appearance. The *Tropheus* sp. from the Yungu area was collected specially for the research for this book, and I would like to take this opportunity to once again thank Siegmund JANICKI and Helmut LÖFFLAD for making the importation possible by taking the majority of the fishes.

Close-up of a male in neutral mood. The delicate pink sheen on the upper half of the head and the green opercular spot are clearly visible. **photo**: P. Schupke

Sub-adults and juveniles have yellowish stripes. The forked tail is already apparent in this sub-adult female. **photo**: P. Schupke

Tropheus sp. South-east; Population 5.8

Trade name: Tropheus "Yellow Chest".

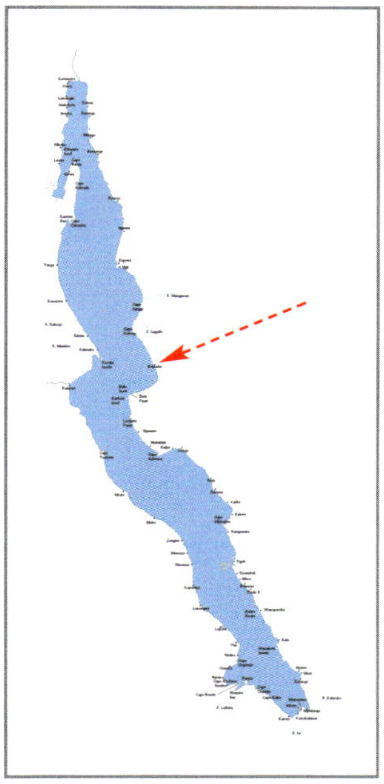

Distribution: The distribution is erranously given as Isonga, but there no *Tropheus* of this population were found there. The actual distribution lies to the north or south of Halembe.

Discovered: This *Tropheus* is first mentioned by KONINGS (Tanganyika Cichlids, 1988) and was discovered by H.W. DIECKHOFF during his numerous diving trips to the lake.

First importation: The first importation took place at the end of the 1980s.

Coloration: The base colour consists of a dark olive brown. Unlike adults, which may lose their yellowish-white vertical barring depending on mood, juveniles exhibit a complete stripe pattern and females more on the upper dorsum. A characteristic feature of this *Tropheus* is an irregularly shaped, brilliant yellow area on the body in the immediate vicinity of the pectoral insertion. The proximal third of the pectoral is likewise yellow. The breadth of the yellow patch on the pectoral fin is rarely more than a centimetre. In much-darkened, sexually active individuals the yellow spot contrasts dramatically with the body. Even sub-adult individuals exhibit the beginnings of a still small yellow spot immediately next to the base of the pectoral. The iris is blue, as in other local forms of lineage 5. The forked tail is unmistakeable.

Remarks: The site at the Tanzanian village of Isonga is a long way from the main distribution. Scientists from the University of Innsbruck were recently in this area, but were unable to find any *Tropheus* matching to the description of *Tropheus* sp. "South-east". As co-author of the book "Tanganyika Secrets" (1992, p. 175, below), Konings mentions a location "near the Malagarasi River delta". It is often the case that inaccurate locality details are given for new discoveries with interesting coloration. The nearest location for this lineage on the Tanzanian coast lies about 100 kilometres to the north at the village of Halembe (Pop. 5.11), and to date no local forms of lineage 5 have been found along this long intervening stretch of coast.

Sub-adult male in normal coloration. The forked tail is already clearly visible even at this age. **photo**: P. Schupke

Tropheus sp. Kabogo; Population 5.9; *Tropheus* sp. Halembe; Population 5.11

Trade name: No trade names are known for either of these local forms.

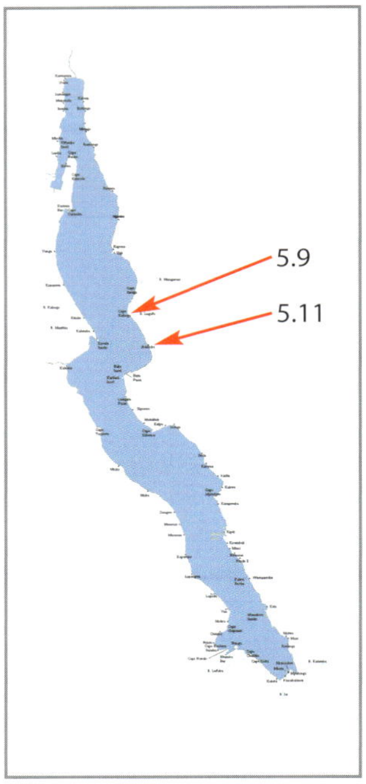

Distribution: Population 5.9 was found at Cape Kabogo; the precise boundaries of its distribution are unknown. Population 5.11 lives in the general area and south of the village of Halembe.

Discovered: A. KONINGS and H. W. DIECKHOFF discovered these *Tropheus* during their diving expeditions in the mid 1980s.

First importation: Neither of these local *Tropheus* forms has as yet been imported for the ornamental fish industry.

Coloration: The base colour in *Tropheus* sp. "Kabogo" is olive green. Depending on mood, the dark colour becomes a lighter greenish beige shade with a yellow cast. Juveniles and females have vertical bars, but adult females and dominant males usually lose their stripes completely, although in fright coloration the stripes may reappear again faintly. The iris is light blue and the caudal forked. There is no yellow lateral spot. *Tropheus* sp. "Halembe" is extremely similar to *Tropheus* sp. "Kabogo" in appearance except that its extreme underside is yellow. The base colour is the same, the iris is the same light blue, and the caudal fin is forked. Again, there is no yellow lateral spot.

5.9

5.11

Remarks: *Tropheus* sp. "Kabogo" contacts the distribution of *Tropheus* sp. "Karago" in the south. The Lugufu river, a few kilometres south of the Malagarasi delta, forms the boundary between these two local forms, and this boundary marks a far from insignificant change in markings. To the south of Cape Kabogo there are further rocky shores, separated by several sandy beaches. Cape Kabogo is also home to *Tropheus duboisi*. Lineage 3 may perhaps also be represented in this region, as *Tropheus* of that lineage are found both to the north and the south. *Tropheus* sp. "Halembe" may also continue further south. It would be interesting to discover why *Tropheus* sp. "Karago", with its attractive lateral spot, is flanked by two *Tropheus* whose appearance is similar apart from lacking the yellow spot (5.10).

The mouths of rivers - in this case the Lufubu - can present insuperable barriers to expansion in *Tropheus* because of the large amounts of sediment they carry into the lake. **photo**: P. Schupke

Tropheus sp. Karago; Population 5.10

Trade namen: Tropheus "Malagarasi", Green Moorii, Tropheus "Lugufu".

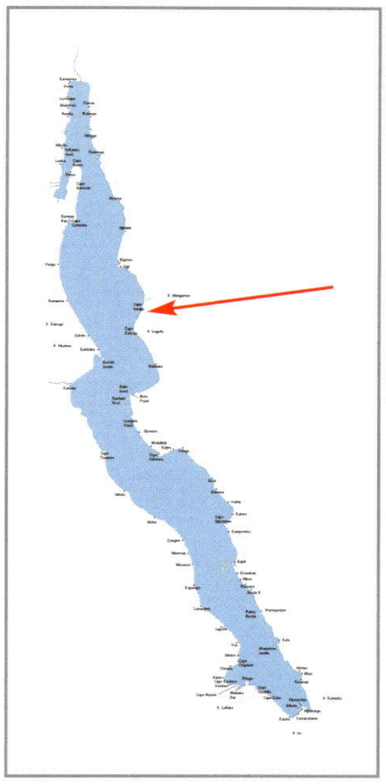

Distribution: The distribution begins to the south of the Malagarasi delta and ends only a few kilometres further south, at the mouth of the Lugufu river.

Discovered: *Tropheus* sp. "Karago" was discovered in 1974 by Wolfgang Staeck.

First importation: The first *Tropheus* of this form arrived in the European ornamental fish industry as long ago as 1975.

Coloration: The base colour is olive green. Depending on mood, the dark colour pales to green-yellow - this is seen mainly when the fish is captured or in fright coloration. Not far behind the head there is a lemon yellow, finger-wide band, seldom extending down to the belly. This band is very variable in its extent. It is larger and more prominent in males. Some individuals also have a yellowish area on the posterior half of the body. The iris is light blue and the caudal deeply forked. Juveniles and sub-adults have yellowish-white vertical barring on the body, but there is little to be seen of this barring in adult specimens, although traces of it may still be visible. All the fins are the same colour as the body.

Remarks: *Tropheus* sp. "Karago" was imported very early on, at a time when only a very small number of species from Lake Tanganyika were available. At that time there was very little information on the lake available, and virtually nothing was known about correct maintenance. And losses were correspondingly high. To some extent these bad experiences with this *Tropheus* must have been passed down over the years, as further imports were rare and this form turned up only very sporadically in the trade. Within its range, *T.* sp. "Karago" is sympatric with *Tropheus duboisi*. Nothing is known of any other species, though it is very probable that *Tropheus* of lineage 3 may be discovered, as they occur to the south at Halembe (Pop. 3. 9) and north of the Malagarasi delta. "Karago" are attractive *Tropheus*, but need to be kept only in large tanks and in adequate numbers, as, if the population is too low, inevitably individual males or large females will become strongly territorial to the detriment of the rest.

Tropheus sp. Kigoma II; Population 5.12

Trade name: None known.

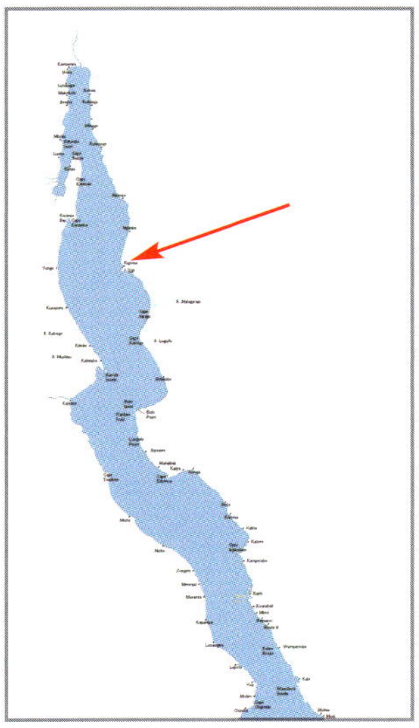

Distribution: The distribution begins to the north of Kigoma and extends northwards along the Ngombe Mountains.

Discovered: Probably by Pierre Brichard and his collecting team at the beginning of the 1980s.

First importation: To the best of my knowledge this form has not been imported for the ornamental fish industry.

Coloration: The base colour of this local form consists of an olive brown, which may, however, become lighter according to mood and then assumes a greenish-grey shade, which may be overlaid with a faint yellowish cast. There is a yellowish-green band, variable in extent, on the starting upper anterior back and generally extending only to the centre of the flank. As a rule adult males no longer exhibit any stripes. All the fins are the colour of the body. Like all local forms of lineage 5, this *Tropheus* has a light blue iris. The caudal is deeply forked in adults with undamaged finnage.

Remarks: As far as I know, this *Tropheus* has hitherto not been regarded as a discrete form. Brichard speaks of the *Tropheus* at Kigoma as being a single species of extraordinary variability. He mentions *Tropheus* with a whitish, yellowish, or light blue iris, body colour varying from green to brown, and light yellow, slate grey, or light green stripes. Lineages 3 and 5 are both found here, as on other stretches of coast. *Tropheus* of lineage 3 have a whitish-yellow iris, while that of *Tropheus* of lineage 5 is light blue. Over the course of time, both lineages have adapted to local conditions and are to some extent similar in coloration. In the aquarium the difference is very quickly apparent. Perhaps the two *Tropheus* have different habitat preferences, as otherwise they would be in competition for food and territory, and such a state of affairs could certainly not have continued for so long. It would also be interesting to know which species occurs in the greater numbers, as actual differences in population density would enable us to draw conclusions regarding the adaptability of the two species.

The fish in the photo to the left is said to have been caught in the area of lineage 5 on the Tanzanian coast. It exhibits a close correspondence to specimens of population 5.10.

photo: H. J. Mayland

Tropheus sp. Nyanza III; Population 5.13

Trade name: Tropheus "Yellow Green"

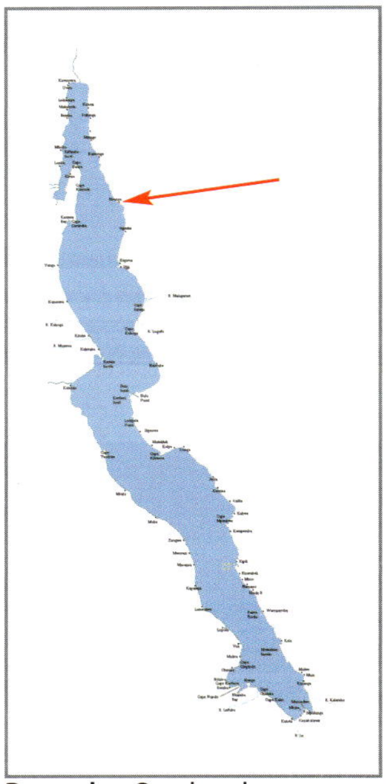

Distribution: The distribution known to date is said to begin north of Nyanza Lac and extend a few kilometres further north.

Discovered: This local form of lineage 5 was discovered and documented by MATTHES in 1962.

First importation: This *Tropheus* has been imported several times, but only during the 1970s.

Coloration: DThe base colour of the body is a dark brown-olive, which can become much lighter according to mood. These fishes then appear more greenish grey-brown with a slight yellowish cast. Adults no longer have stripes. A green band, of very variable extent, begins beneath the start of the dorsal and may extend down to the belly region. There may also be iridescent greenish spots on the gill-covers. In some individuals the greenish band becomes lighter and yellow shades become visible. The iris is brilliant light blue. Juveniles and females have emerald green vertical stripes on a brown background.

Remarks: On the shore opposite Nyanza Lac there are *Tropheus* of lineage 5 which also exhibit emerald green shades, namely population 5.3 at Katende. If we look at the various depths shown on the map of the lake then we can see why these two local forms are somewhat similar in coloration. When the water level was low the two populations were immediately adjacent. On the coast north of Nyanza Lac the *Tropheus* of this local form encounter other lineages. Firstly there is the *Tropheus* of lineage 3 (population 3.13), known as the striped Kigoma *Tropheus* Then there is the *Tropheus* of population number 2.1, the Chocolate Moorii, which was described as *Tropheus brichardi* in 1975. These forms can be confused, not only in the lake but also in the aquarium when the fishes have not yet coloured up. BRICHARD states that on both sides of a single bay there may be not only normal Chocolate Moorii (here, pop. 2.1) but also a variant with alternating chocolate brown and splendid emerald green stripes (here, pop. 5.13). In addition there are also *Tropheus* of lineage 3, which have yellowish-white stripes on a brownish background, plus a light-coloured dorsal. Absolute chaos!

Tropheus sp. Rumonge II; Population 5.14

Trade name: None known.

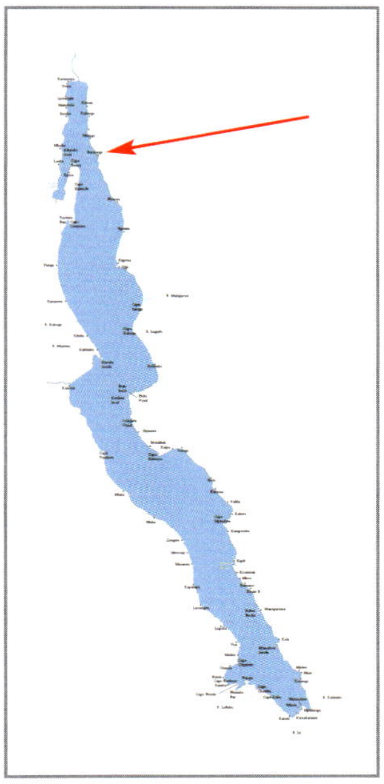

Distribution: The known distribution at present is said to be a few kilometres north of the village of Rumonge, which lies in the southern part of Burundi, diagonally north-east of the northern end of the Ubwari peninsula.

Discovered: There is no record of when this *Tropheus* was discovered. To the best of my knowledge, it is first mentioned by A. KONINGS in his 1988 book, "Tanganyika Cichlids".

First importation: As far as I know there have not yet been any commercial importations of this local form.

Coloration: The base colour of this *Tropheus* is said to be brown-olive. There is a green-yellow band, or variable intensity, on the anterior dorsum. By and large this *Tropheus* is very similar to the more southerly populations at Nyanza Lac (Pop. 5.13) and Kigoma (Pop. 5.12).

Remarks: The external appearance under water is similar to that of *Tropheus* sp. "Rumonge I" of lineage 1 (Pop. 1.3) and they may be difficult to differentiate in murky water. Both *Tropheus* have a lighter (compared to base colour) spot on the upper anterior dorsum. In addition, the base colour is almost the same. There is no doubt that the small differences would. not be noticeable on superficial examination underwater. As far as I know, neither of these *Tropheus* has as yet been imported for the ornamental fish industry. Their rather monotonous coloration makes them of little or no interest to aquarists.

Tropheus of lineage 6

The *Tropheus* of this lineage have a number of peculiarities in their coloration, markings, and distribution, so there appears to be justification for assigning them to a lineage of their own. All four populations share a common grey-brown-beige base colour with a bluish-grey to brownish-grey dorsal, anal, and ventrals. There is a striking stripe pattern on the forehead. In addition, three populations have a more or less light spot in the middle of the back, immediately beneath the dorsal, which may disappear depending on mood. The caudal is normally indented. The distribution of the populations is rather scattered, interspersed with those of other *Tropheus* lineages. The populations in Tanzania, at Cape Kibwesa and a few kilometres to the east raise a number of questions. How they got there remained a mystery for a long time. If we look at the map showing the underwater contours of Lake Tanganyika, the matter becomes less of a puzzle. As the water level rose again and the separate lakes became one, then as the water rose and the isthmus between the two lakes became increasingly smaller, *Tropheus* of population 6.3 may have migrated diagonally across the lake from their original home north of Moba in the former Zaire. But the reverse is also conceivable. In which case, then as the lake level dropped they may have migrated away from their original home, which would then have been near Cape Kibwesa, along the east coast to where they encountered other *Tropheus* and so had to continue their wanderings. These hypotheses sound quite plausible, but they are only theories and can be confirmed or disproved only with the aid of genetic studies. Perhaps following such studies it will be possible to say whether this lineage once had a wider distribution. It may well be that the four current populations are relicts from more prosperous times.

The *Tropheus* of lineage 6 come into contact with other lineages within their distribution area. The local form at Kapampa (Pop. 6.1) encounters *Tropheus* of lineage 7 at Kiku, to the north of Luhangwa in the southern Congo. According to Christian HOULLIER, in "The Cichlids Yearbook", volume 4, edited by Ad KONINGS, the two forms are sympatric there. North of Kapampa population there are *Tropheus* of another lineage, which live north and south of the village of Mwerazi and likewise appear to be somewhat isolated. They belong to lineage 8, the Rainbow *Tropheus*. It is not known whether they contact *Tropheus* sp. "Kapampa". To the north of the Mwerazi population there are again *Tropheus* of lineage 6, living north of the village of Zongwe. It is not known how far further north their distribution extends. At Cape Kibwesa *Tropheus* sp. "Kirschfleck" (lineage 12) and *T. polli* contact *Tropheus* sp. "Kibwesa". In the summer of 2001 a further population, very similar to that at Cape Kibwesa, was discovered by Helmut LÖFFLAD to the east of the Cape. The western shoreline has been explored only here and there, and precise locality details are not available for the areas in the Congo. For this reason the information cited should be treated with circumspection. Lineage 6 contains the populations numbered 6.1, - 6.4.

Distribution of lineage 6

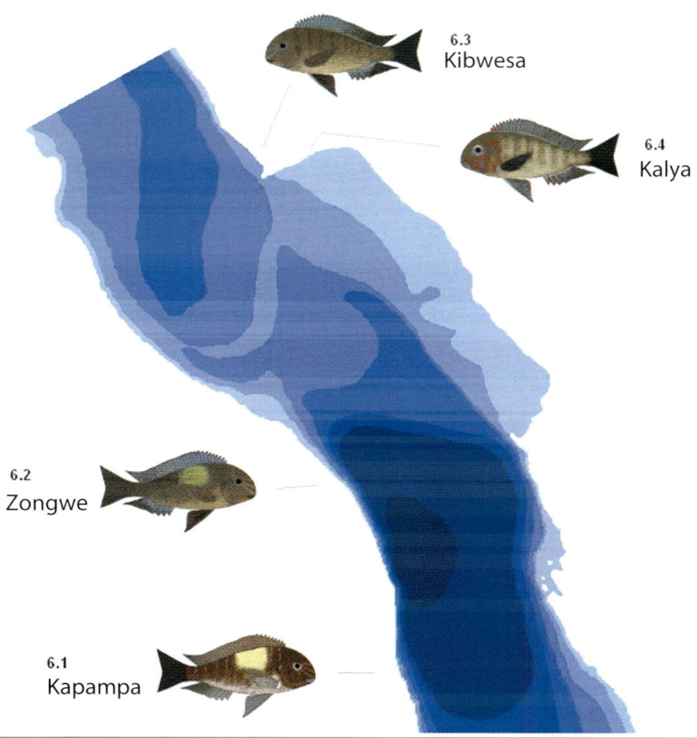

Tropheus sp. Kapampa; Population 6.1

Trade name: Tropheus "Kapampa".

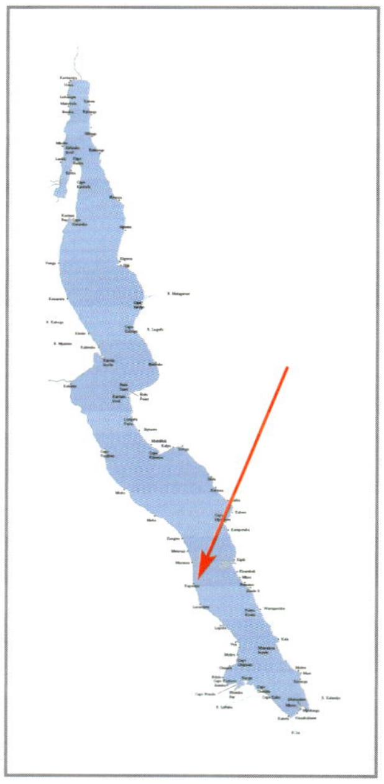

Distribution: The distribution begins at Kiku, a few kilometres north of Lunangwa, and continues - as far as is known - to Kapampa Bay .

Discovered: This local form may have been discovered by aquarists. I do not have any more precise information.

First importation: The first importation for the ornamental fish industry was by S. JANICKI in 1974 or 1975.

Coloration: The base colour consists of a dark brown, occasionally overlain with an aubergine-coloured cast. The body scales are dark only at their base, and the posterior part of each scale is more or less light, creating a reticulated pattern on the body. The entire lower part of the body is much lighter. On the mid-flank there is a usually ivory-coloured spot, which may, however, appear yellowish-green. The spot starts immediately beneath the dorsal and ends on the central flank, extending horizontally across 5-8 scales. The mouth and throat are brilliant turquoise blue. There are two bars across the nose, linking the eyes. A third stripe runs from the bottom of the eye to the corner of the mouth. There are scattered light blue dots on the head. The cheeks have a dark pink cast. Depending on mood, the light spot may disappear. Traces of the juvenile barring may remain here and there beneath the dorsal in adult fishes.

Remarks: At present we can only speculate regarding the origin of these *Tropheus*. However, in view of the external appearance of this lineage, I regard it as rather unlikely that these *Tropheus* belong to the Rainbow *Tropheus* (lineage 8) as suggested by Ad KONINGS. In the south of their range they encounter *Tropheus* of lineage 7, and, according to KONINGS, are sympatric for a few hundred metres. These two lineages certainly do not occupy the same ecological niche and so it is highly probable that they are found together in various places. Yet another point that requires clarification. Studies of this population are so far lacking, and would doubtless clarify the phylogenetic position of these fishes.

Tropheus sp. Zongwe; Population 6.2

Trade name: Tropheus "Zongwe".

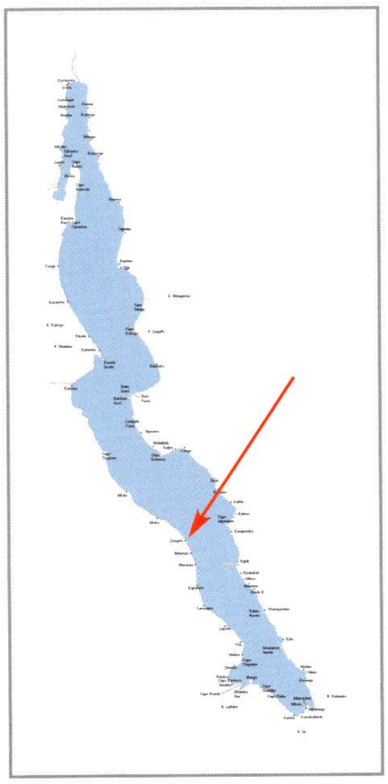

Distribution: The distribution begins to the north of the village of Zongwe. Its northern limit remains unknown.

Discovered: This local form was probably discovered by Pierre BRICHARD and his collecting team in 1986.

First importation: This *Tropheus* was first imported for the ornamental fish industry in 1986.

Coloration: The base colour consists of a grey-beige-brown. Depending on mood the colour can become much darker, and the fish then appears almost black with a greenish cast. There is a yellowish to greenish lateral spot, very striking against the dark body, beneath the centre of the dorsal. This spot extends horizontally across 7-9 scales, and vertically to round about the centre of the flank. In passive individuals the underside is very light. Dorsal and anal have a bluish sheen, and the anal has a few small orange spots on its spinous part. The caudal is dark, and, like the ventrals, inconspicuously coloured. Mouth, cheeks, and breast are a pale pink colour. The three lighter stripes on the forehead are very striking. Immediately below the operculum there is an orange-red area, extending to behind the pectoral insertion. Sometimes, if the fish is in a fairly passive mood and relatively light in colour, there is a dark pink cast all over the body, as in *Tropheus* sp. "Kapampa".

Remarks: Unfortunately, the available information on the distribution is very sparse. It is, however, possible that *Tropheus* sp. "Zongwe" comes into contact with lineage 8, whose range lies to both the north and the south. A note in the research of STURMBAUER and MEYER states that they caught a *Tropheus* with only four anal spines at Zongwe. This may have been a *Tropheus* of lineage 4 (Pop. 4.3). The *Tropheus* from Zongwe have seldom been imported, and hence aquarium observations of the behaviour of this local form are correspondingly rare. Unfortunately the current situation in the Congo is less than propitious, as there are once again political problems that make it impracticable to collect fishes there. *Tropheus* have been imported as *T.* "Zongwe" that were actually northern "Rainbows" (lineage 8) and very similar to *T.* sp. "Murago".

A deceptively peaceful evening scene. Anyone who goes fishing at Cape Chipimbi (the tongue of land in the background), which forms the border between Zambia and the Democratic Republic of the Congo, is putting his life at risk. **photo:** P. Schupke

Tropheus sp. Kibwesa; Population 6.3

Trade namen: No trade name is known.

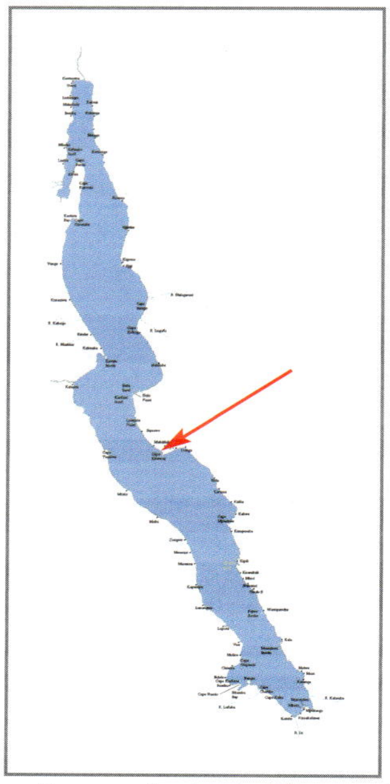

Distribution: As far as is known at present, the distribution is limited to Cape Kibwesa.

Discovered: This local form was first mentioned by Konings in his 1988 book "Tanganyika Cichlids".

First importation: As far as I know this form has never been imported for the ornamental fish industry. It may have been brought back by aquarists, but has not yet shown up in the trade.

Coloration: The brown-grey base colour is unbroken by any spot. Only the underside is much lighter and vaguely reminiscent of the Kalambo and Mutonwe *Tropheus* of lineage 8 in southern Zambia. The caudal is dark, as are the posterior parts of the dorsal and anal. The rest of the dorsal, and the spinous part of the anal, are bluish in colour. The base of the anal and the three stripes on the nose are a striking light grey-blue. The pectorals and ventrals are dark and virtually match the posterior dorsal in colour. The leading edge of the ventrals is a light grey-blue colour. Because the base of each scale is darker than the posterior part, there is a slight reticulated pattern on the body. The caudal is normally indented.

Remarks: *Tropheus* sp. "Kibwesa" is one of the very few *Tropheus* for which, as far as I know, there are no aquarium data available. There is apparently only one photo of this fish in existence. To the north its distribution adjoins that of lineage 12, *Tropheus* sp. "Kirschfleck". *Tropheus polli* has also recently been discovered there, but is the only form to occur past Cape Kibwesa. Genetic research on *T.* sp "Kibwesa" by scientists at the University of Innsbruck has shown that they derive from the ancestral lineage from which lineages 7 and 8 also originate. *T.* sp. Kalya (Pop. 6.4), recently imported for the first time, is possibly very closely related to *Tropheus* sp. "Kibwesa".

Tropheus sp. Kalya; Population 6.4

Trade namen: Tropheus "Njano".

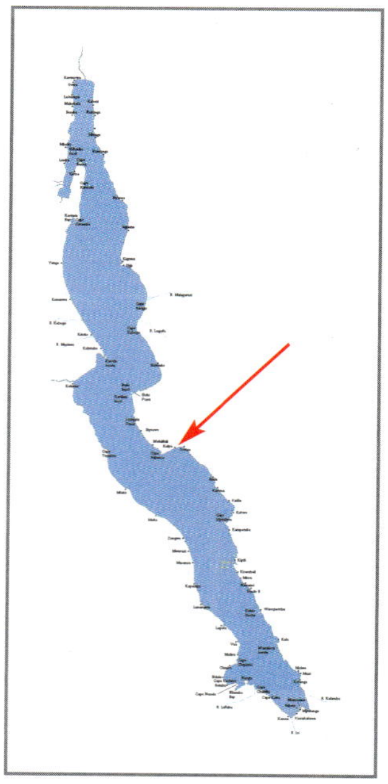

Distribution: To date known only from Kalya, which lies between Cape Kibwesa and Isonga.

Discovered: This *Tropheus* was accidentally discovered by Helmut Löfflad of Bühl, when he stopped to buy fruit at the nearby village.

First importation: Summer 2001.

Coloration: The head is generally a medium brown colour with a reddish cast, and can look pink, especially in the eye region. Some individuals have a light area, reminiscent of a triangle in shape, below the eye on the anterior part of the operculum. The forehead and anterior dorsum are the same colour as the head. The rest of the dorsum, as far as the tail, is greenish. The flanks and the belly are whitish-cream and may be crossed by up to six rather darker vertical stripes. The dorsal is blue-grey and may exhibit small reddish-brown spots, mainly on the anterior part of the fin and at its base. The base of the pectoral is noticeably pink. The caudal is almost black. The upper half of the body may become much darker, depending on mood, and the white vertical bars are then very prominent.

Remarks: The external appearance of this population resembles that of *Tropheus* of lineage 6, but also some forms of lineage 8. The very close proximity (only a few kilometres) to *Tropheus* sp. "Kibwesa" suggests that *Tropheus* sp. "Kalya" belongs to lineage 6. Only DNA research can clarify the situation. The discovery of this colour form demonstrates yet again that not all *Tropheus* populations are known by a long chalk. Because of its plain coloration this colour variant will probably make little impression on the aquarium hobby, though it is certainly of interest to *Tropheus* freaks.

Long live the lunch break, which was responsible for the discovery of *Tropheus* sp. "Kalya". **photo:** P. Schupke

Tropheus of lineage 7

The local forms of this lineage have a number of peculiarities which clearly differentiate them from *Tropheus* of other lineages. These *Tropheus* are unique in both their coloration and their geographical range. Their distribution begins near to Kiku, which lies only a few kilometres north of the Lunangwa in the southern Congo, and ends at the Nangu peninsula, just into Zambia. The coloration of lineage 7 is dependent on the locality of each form. The base colour of lineage 7 varies from form to form, progressively from north to south throughout its distribution, ranging from brown-beige through brown to red-brown, then lightening to brown-orange through golden brown to yellowish. In neutral coloration, the dorsum of southern local forms is dark to light green. The only common feature is the bars on the nose, which are not as prominent as in lineage 6 further to the north. Adult specimens usually show no or only faint traces of their juvenile barring. They have six anal spines. As regards the fins, only the dorsal, anal, and ventrals are brightly coloured. The caudal is only slightly indented and, along with the pectorals, uniformly dark. The entire lineage exhibits a red shade in the dorsal and/or anal, which varies in intensity and the amount of colour according to the locality. In the northern part of the lineage we find a dark blood-red, which, the further south we go, noticeably changes to a light red or orange and then to yellow and yellow-white. The light coloration of the individual local forms is connected with the locality where they are captured. In the north - with the exception of Kapampa Bay - the rocky areas underwater are dark, unlike the lighter shores further south which are characterised by shallower water and lots of sand. In Nkamba Bay this changes again, and the *Tropheus* living there are rather darker. Juveniles have light vertical bars, which are noticeably wider than in the Rainbow *Tropheus* of lineage 8 further to the south. Moreover the stripes disappear appreciably sooner than in the *Tropheus* of lineage 8. To the discomfiture of aquarists, dominant *Tropheus* unfortunately become much darker and then there is little to be seen of their once attractive colours. They look their best in undecorated and brightly-lit aquaria or in white buckets - both types of container have proved very effective marketing tools in the past. At the northern extreme of their range these *Tropheus* encounter *Tropheus* of lineage 6, with which they are supposedly found sympatric in the vicinity of the village of Kiku. In the south the meet at the end of Nkamba Bay *Tropheus* sp. "Nangu", which belong to lineage 8. I would like to share an interesting observation from my fish-room with you, dear reader. A bought-in pair of *Tropheus* sp. "Ilangi" finally produced fry after a long wait. Of course, I was absolutely delighted when the eggs developed and a furtive glimpse into the slightly open mouth after several weeks revealed tiny *Tropheus* inside. And then came the big surprise. Of the nine well developed fry, a third were lighter than the rest. The parents had not been kept with other *Tropheus*. In the course of the weeks that followed, the "wrong" coloration of the three youngsters became increasingly more apparent compared to the other six (Photos on p. 86), which developed into attractive, true "Ilangis". The light-coloured individuals looked very much like the "Yellow Rainbow". So, how can this variation in the young "Ilangis" be explained? Was it a spontaneous mutation, or had the ancestors of my adult individuals had contact with the "Yellow Rainbow" *Tropheus*? I don´t know.

During the 1980s all these local forms were regularly imported. Then Adrian Karr´s export company at Ndole Bay closed down, and the *Tropheus* of this lineage were rather less frequently imported. There is now a new company operating between Ndole Bay and Cape Kachese, so the attractive *Tropheus* from this region are again being exported more regularly.

At present up to 15 populations are recognised in this lineage.

Distribution of lineage 7

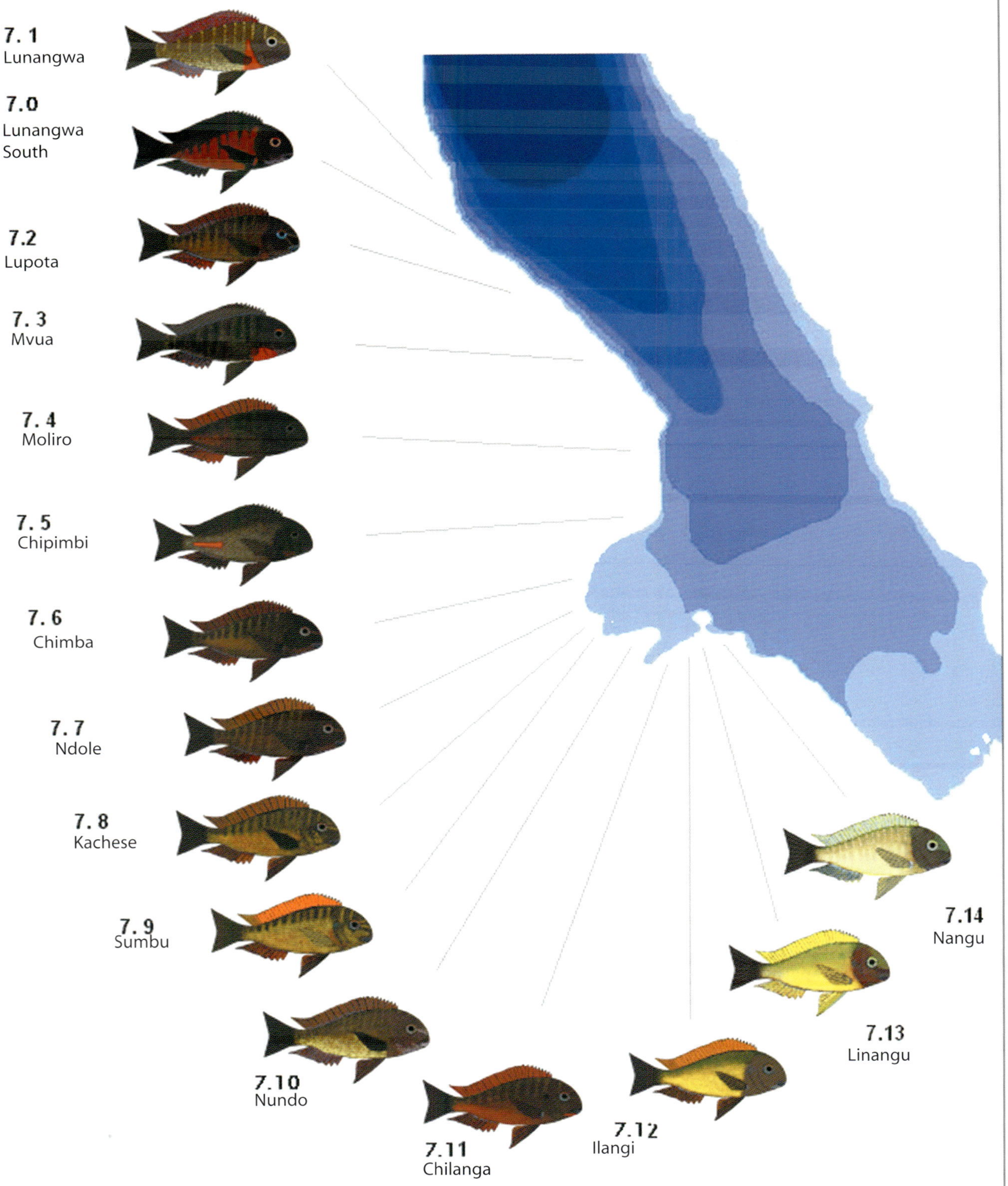

Tropheus sp. Lunangwa South; Population 7.0

Trade name: No trade name known.

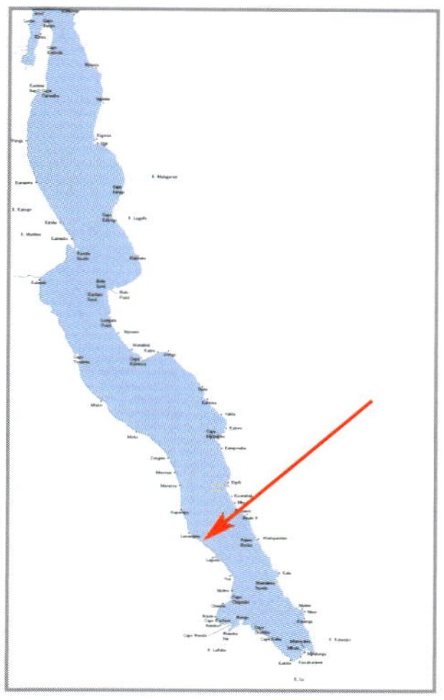

Distribution: The name of the population refers to the collecting site south of the Lunangwa on the Congolese coast.

Sonstiges: The only information on these fishes originates from the Internet. I found this fish on the home page of the Swedish company "African Diving". This *Tropheus* appears to me, judging by the photo published on the home page, to be a discrete population. Unfortunately further information on these fishes is not at present available.

Tropheus sp. Lunangwa; Population 7.1

Trade name: Tropheus "Kiku".

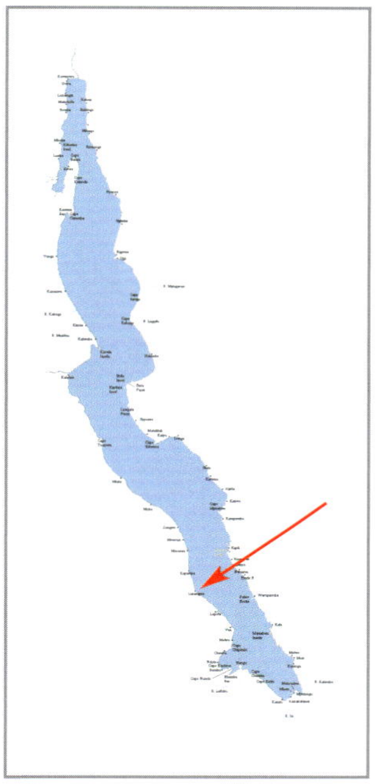

Distribution: The distribution is restricted to a 20 kilometre section of coast south of Kapampa Bay.

Discovered: This local form was probably discovered by Pierre BRICHARD and his collecting team in the mid 1980s.

First importation: The first importation of small numbers took place at the beginning of the 1990s.

Coloration: In individuals south of Kapampa Bay the base colour consists of a beige-brownish shade, which may be rather greenish on the back. Further south the base colour becomes more reddish. The light vertical bars of juveniles are almost completely lost by adult and sexually active males. Only in females and juveniles are they constantly apparent. The mouth, throat, and base of the anal are a light violet-blue colour. Dorsal, anal, and ventrals are dark with a light reddish admixture, which may sometimes be overlain with light violet in places. The ventrals are plain dark in the pic, quite unlike the D and A. The anal has numerous blood-red dots. Behind the head there is a narrow reddish brown band in the north, and further south a red band behind the gills and on the breast. Territorial individuals become much darker.

Remarks: The coast south of Kapampa Bay is characterised by a high degree of sedimentation, which has clearly had an effect on the base colour of the *Tropheus* living there. The beige-brownish to beige-orange shades are very similar to the sediment in colour, and probably offer enhanced protection against predation. The less the amounts of sediment, the darker the *Tropheus* become. This phenomenon is repeated several times at other sections of shore within the distribution of this lineage. Because of their relatively unattractive coloration the *Tropheus* of this local form have unfortunately found few fans. In addition, the political situation means that it is impossible to travel in this area without risking life and limb. The border zone between the Congo and Zambia is likewise not recommended, as the border seems to be re-drawn every day according to whim. We made a short excursion to Moliro virtually without problem. but an acquaintance who went only as far as Cape Chipimbi was intercepted by a patrol boat and taken away at gunpoint. Up to that point he had no idea he was going to spend a six week unplanned "holiday" in Kalemie.....

The colour differences between these two *Tropheus* can be clearly seen. The lower fish was caught south of Kapampa Bay and the other near to the mouth of the Lunangwa. **photo**: P. Schupke

Tropheus sp. Lupota; Population 7.2

Trade name: Tropheus "Lupota", Red-Breasted Moorii.

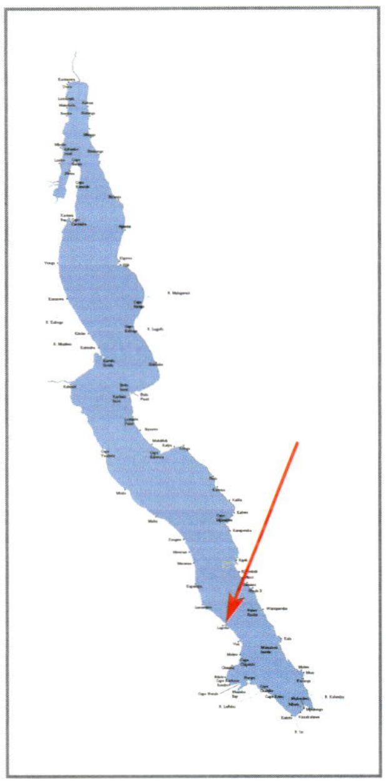

Distribution: The distribution of this local form begins a few kilometres south of Lunangwa Bay and continues to Lupota Bay.

Discovered: No date known.

First importation: First importation: This local form was first imported for the ornamental fish industry in 1985.

Coloration: The base colour consists of a red-brown, with a greenish cast on the dorsum. Mouth, chin, and the base of the anal are clear light violet to light-bluish. The breast region and the pectoral insertion are vivid blood-red. The light vertical barring of juveniles and females almost completely disappears in males and when sexually active they become uniformly dark with no barring at all. Dorsal, anal, and ventrals exhibit a dark red shade, which may be more of an iridescent light violet at the dorsal base. Traces of the bars may also be visible at the base of the dorsal. The anal is decorated with numerous rather elongate, dark red dots. The caudal and pectorals are dark in colour. As in all local forms of this lineage, the forehead has three lighter stripes running from eye to eye. The iris is blue with a dark red area in its upper part.

Remarks: The red colour of these *Tropheus* is very mood-dependent. In well-lit bare tanks this form exhibits an attractive red and the dark pigmentation is completely suppressed, but this fright coloration is never or only rarely exhibited in a well-decorated aquarium. *Tropheus* sp. "Lupota" is a relatively plain form of the *Tropheus* found in the Moliro region. This form was more popular during the 1980s than today - it is currently maintained only rarely, if at all. It cannot compete with the appreciably more attractive colour variants that have been imported in recent years.

Tropheus sp. Mvua; Population 7.3

Trade name: Tropheus "Vua"

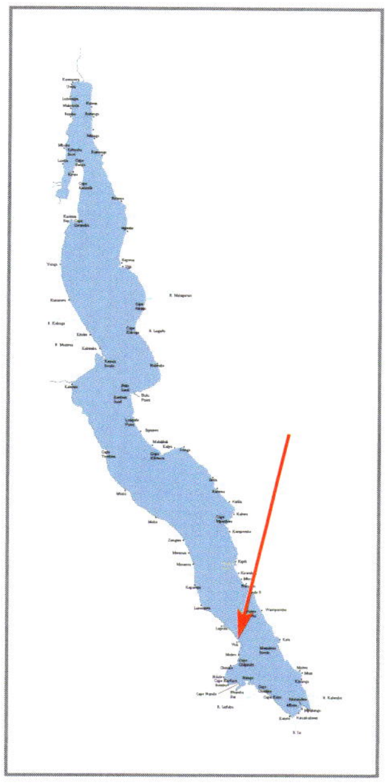

Distribution: The distribution lies to the north of Mvua Bay and extends to Lupota Bay.

Discovered: This local form was discovered by SCHEUERMANN and STAECK in the mid 1970s.

First importation: *Tropheus* "Vua" was probably imported in small quantities soon after its discovery. A few years later it was imported in larger numbers.

Coloration: The base colour consists of a dark brown. In this local form the throat region, the breast, and sometimes even the lower part of the gill-cover is a striking blood-red. The rest of the body and the fins lack any red component as base coloration, but the anal fin has red spots. The iris is slightly bluish, with red in the upper part. Adult specimens normally cease to display any vertical barring, although this may reappear in unfamiliar surroundings. Apart from the red-spotted anal the fins are an inconspicuous brown-grey. The forehead has light grey-brown stripes. Sexually active individuals become very dark.

Remarks: The extent of the red patch can very considerably from individual to individual. in some specimens only the throat is red, in others just the breast area. In particularly beautiful specimens the breast, throat, and the lower part of the gill-cover is a vivid blood-red colour. Males often exhibit more colour than females. It is probable that this local form no longer exists in captivity, as quite early on more colourful forms of various lineages out-competed it for space in out aquaria. As far as I know, few, if any, individuals of this population were imported during the early 1990s. More recently (December 1998) *Tropheus* from this region appeared once again, and included, inter alia, some *T.* sp. "Mvua" The much more colourful *Tropheus* from Moliro are more popular with aquarists and correspondingly more frequently ordered by importers. According to the researches of Christian STURMBAUER and his colleagues, *Tropheus* sp. "Mvua" is phylogenetically close to the *Tropheus* from Chaitika and Wapembe, as well as to the "Kirschfleck" from north of Cape Kibwesa.

Tropheus sp. Moliro; Population 7.4

Trade name: Tropheus "Moliro".

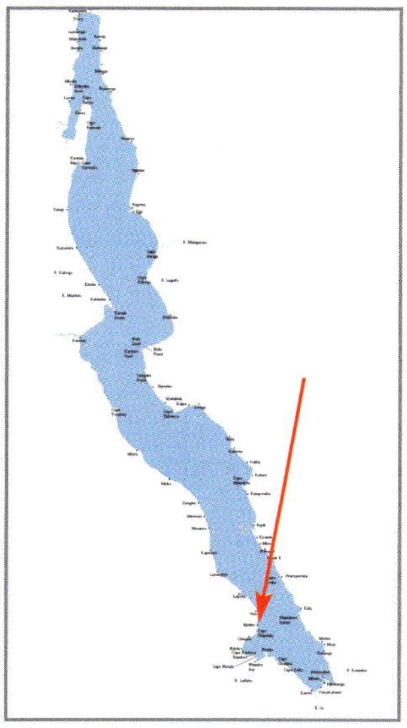

Distribution: This local form lives in the vicinity of the village of Moliro in the extreme south of the Congo. The precise extent of its range is unknown.

Discovered: This *Tropheus* was discovered by POLL during his major expedition to Lake Tanganyika in 1946-47.

First importation: This colour form was imported for the ornamental fish industry as long ago as 1976.

Coloration : The base colour consists of a red-brown, which, depending on mood, can become much darker or much lighter. The brightest red is seen in fright coloration. The red colour may be more prominent on the posterior part of the body. Adult and sexually active males generally no longer exhibit any barring. Mouth, throat, pectoral insertion, and anal base are slightly bluish-grey in colour. The upper dorsum is somewhat greenish. There are three lighter stripes on the nose/forehead region.

A characteristic feature of the Moliro Tropheus is the intensely red-coloured dorsal. Every other spine of the dorsal is darker. The ventrals often exhibit the same intensity of colour. Caudal and pectorals are very dark, virtually black.

Remarks: Import consignments regularly include individuals that are very similar to the neighbouring population further south at Cape Chipimbi. These fishes show hints of a red bar on the posterior part of the body, as is usual in the Tropheus from Cape Chipimbi. These specimens are also noticeably lighter and have rather less red on the rest of the body than in the Moliro. It often happens that intermediate forms exist between local forms, and then the characteristic coloration occurs only along a particular section of coast. For this reason, imported consignments are regularly also variable, which does not make life easy for the hobbyist and breeder who wish to buy more fishes to add to an existing group. But the intermediates can also be very attractive in their coloration (see photo below).

This *Tropheus* exhibits colour characters of *Tropheus* from both Moliro and Chipimbi. **photo**: P. Schupke

Tropheus sp. Chipimbi; Population 7.5

Trade name: Tropheus "Chipimbi".

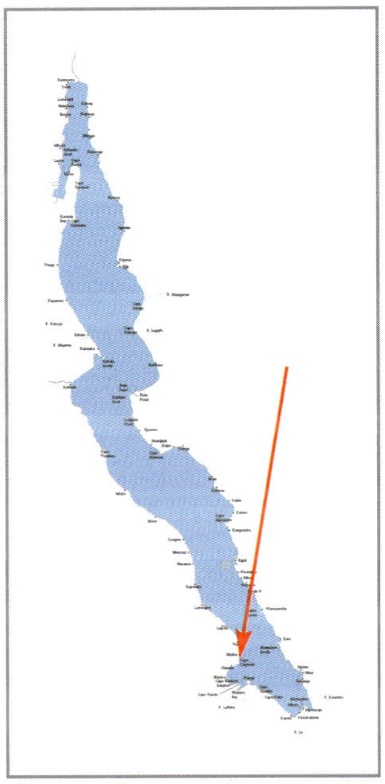

Distribution: *Tropheus* of this colour form are caught at Cape Chipimbi, which lies on the border between the Congo and Zambia.

Discovered: Poll discovered this *Tropheus* during his hydrobiological expedition in 1946-47.

First importation: These *Tropheus* first entered the ornamental fish industry in 1975, but were not found regularly in importers´ tanks until the beginning of the 1980s.

Coloration: The base colour of the body is a light whitish grey-brown, with dark at the base of each scale. The head is dark brown with three faintly visible stripes on the forehead. Adults do not normally continue to exhibit vertical barring. Mouth, throat, the start of the anal and the base of the dorsal are vivid light blue. The caudal has blue-grey areas on the membranes but is otherwise dark. The dark pectorals stand out boldly against the light red body. The iris is red-orange to red in colour. Adult specimens may become very light, almost whitish, on the body. This local form exhibits a red area, variable in both intensity and extent, on the posterior body; this may be a narrow, intense red narrow stripe, or the red may be rather less intense but the coloured area larger. The red chin spot seen in a previously published photo is actually very rare in this colour variant and is normally a feature of the Mtossi local form.

Remarks: Among the Chipimbi *Tropheus* too we find aberrant individuals that are similar to the *Tropheus* at Moliro. Some specimens of the Chipimbi colour form are very attractive. Intermediate forms occur now and then in import consignments. The same occurs all too often when local forms are kept together (eg out of ignorance). Interestingly, the two forms exhibit completely different fright coloration. The *Tropheus* from Cape Chipimbi lose all their dark pigment and look rather washed-out, while the *Tropheus* from Moliro, on the other hand, do not become anything like as pale. This is readily apparent if the two are compared (see photo below). Juveniles are usually easily distinguished by the absence of reddish coloration in the dorsal. In the southern part of this colour form´s range the red stripes become narrower and less intense, and - the most striking feature - the red area on the posterior body may be fragmented into two or three short streaks.

In fright coloration *Tropheus* sp. "Moliro" (behind) and *Tropheus* sp. "Chipimbi" can readily be distinguished from each other.
photo: P. Schupke

Tropheus sp. Chimba; Population 7.6

Trade name: Tropheus "Chimba", Red-Black Moorii, Zambian Red.

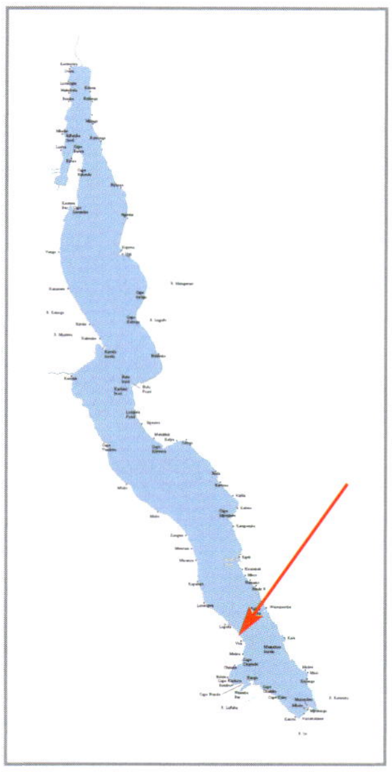

Distribution: This colour form is collected in the general area around Chimba in Cameron Bay, Zambia.

Discovered: This *Tropheus* was discovered in the mid 1970s by German aquarists, and documented by STAECK and SCHEUERMANN in 1975.

First importation: The Red-Black Moorii was first imported in the mid 1970s. More regular imports followed during the 1980s.

Coloration: The base colour consists of a brown-red, which, depending on mood, may become more dark brown (in motivated individuals) or uniform red (fright coloration). Dorsal, anal, and chin are a lovely carmine red. The posterior base of the dorsal, thee anal, and the pectorals are bright violet-blue. The upper part of the dorsum looks greenish. Adult usually no longer exhibit barring. There may be a reddish area, whose edges are far from clearly delineated, on the posterior body. In highly aroused individuals red areas appear between the barely visible stripes on the forehead. Juvenile *Tropheus* sp. "Chimba" are attractive: a red reticulation covers the entire body, which is green above the lateral line and beige below. Very young individuals, 2-3 cm in length, do not as yet have any red on the body, but are brown-black with grey-whitish vertical bars. The bars continue into the lower part of the dorsal. The iris has a bright yellow inner edging; that apart, the eye is bluish-silvery except for the reddish upper part.

Remarks: The Chimba *Tropheus* was sold as the "Coral Red Moorii" in Germany at the end of the 1970s. So why is that not included in the trade names? MB These fishes looked truly impressive in appearance, at least in tanks with very light sides and bottom. For practical reasons, of course, these tanks were undecorated. It was known even then how best to present certain *Tropheus* colour forms. In the southern part of its range the red colour gradually changes to an ochre-orange. *Tropheus* sp. "Chimba" is very similar in appearance to *Tropheus* sp. "Chilanga" (Pop. 7.11) from the south-eastern part of Nkamba Bay, but the latter lacks the vivid light violet colour at the base of the fins, and its red is brighter and more apparent, particularly on the throat, breast, and belly.

Tropheus sp. Ndole; Population 7.7

Trade name: Tropheus "Ndole".

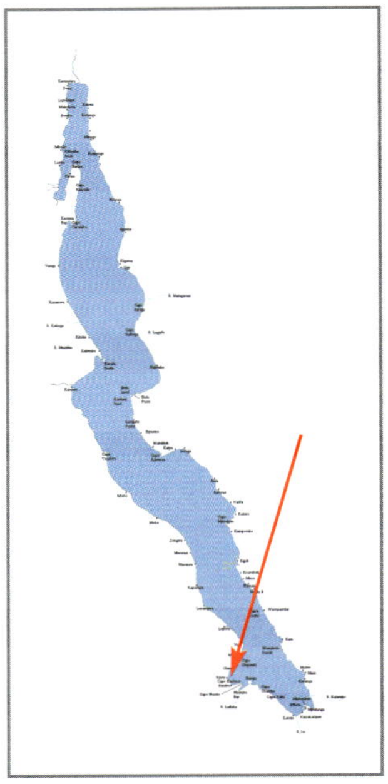

Distribution: The distribution area lies in Ndole Bay and adjoins the distribution of Kachese colour form. It is not known how far north it extends.

Discovered: No details are known regarding the discovery of this form..

First importation: In 1983 a few specimens were collected by aquarists and taken back to Germany. Later on this form was collected for the ornamental fish industry and imported a few times.

Coloration: The base colour consists of a dark ochre-orange, with the head rather darker in colour. There is a faint green hue on the back immediately beneath the dorsal. The entire underside is orange-brown, becoming fainter towards the back. One of the stripes on the forehead is clearly visible, and a second almost as distinct. The dorsal is a vivid red-orange colour, while the anal and ventrals have only a touch of this colour. The posterior part of the dorsal, the base of the anal, and the pectoral insertion are light violet, as in the previous colour forms of this lineage. Every other dorsal spine is darker in the anterior part of the fin, with this coloration extending to every spine further back. The outer edge of the eye is bordered with orange-yellowish, and the inner iris is a bluish-silver colour.

Remarks: It happens now and then that a consignment of Kachese *Tropheus* contains the odd *Tropheus* from Ndole-Bay. The reason for this lies underwater. A large part of Ndole Bay is only about a metre deep with numerous flat sandstone surfaces full of holes eroded by the surf. This cover, which is home to the Ndole *Tropheus*, provides them with a link to the rocky littoral of Cape Kachese. The Ndole *Tropheus* looks like a lighter version of the Chimba *Tropheus* (Pop. 7.6). The red blends in well with the relatively light-coloured environment in Ndole Bay, and over a long period of time this colour character as become fixed in the genetic inheritance of this and numerous other local forms around the lake. There are sometimes small intermediate populations between the main populations of these and other *Tropheus*.

Ndole Bay Lodge in Zambia is ideal for combining an exciting family holiday with a serious collecting tour. **photo:** P. Schupke

Tropheus sp. Kachese; Population 7.8

Trade name: Tropheus "Kachese".

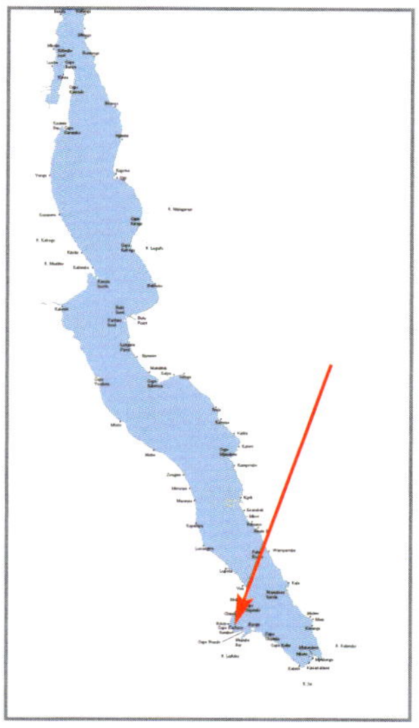

Distribution: The distribution extends from the end of Ndole Bay, round Cape Kachese, and a little way into Sumbu Bay.

Discovered: The *Tropheus* at Cape Kachese were discovered by German aquarists in the mid 1970s.

First importation: The first importation took place as early as the mid 1970s, and this form still regularly appears on importers´ lists today.

Coloration: The base colour is usually divided into two equal halves, divided longitudinally along the centre of the flank. The upper part is a greenish colour, while the lower part is orange-yellow to orange-red. The head is dark and orange-yellow only in the chin area. The dorsal is light red-orange, and the same colour, only paler, is seen in the anal. As in the Ndole *Tropheus*, every second dorsal spine is lined with black. Depending on mood, the orange-yellow colour of the lower body may disappear, giving way to a uniform darker green-ochre. In fright coloration, vertical stripes are seen in adults as well. The base of the posterior dorsal and that of the anal are a light violet colour. There is a single stripe, the colour of the lower body, on the forehead. Caudal and pectorals are very dark.

Remarks: Around Cape Kachese the water is often clouded for long periods, so that nothing much can be made out underwater. And it is far from rare for large crocodiles from the nearby Sumbu National Park to wander by, preventing diving and snorkelling in the area. Some importations include individuals with a strikingly high red component on the lower half of the body. This red may derive Ndole *Tropheus* which now and then come into contact with the Kachese population via the underwater sandstone flats. The black delineation of the dorsal spines is more prominent in these specimens. In the mid 1980s, when there was still a collecting station north of Ndole Bay, *Tropheus* from this part of Zambia were regularly seen. There was even a wonderful lodge where I stayed for three weeks in 1983. In those days Adrian Karr had a collecting station almost on the doorstep.

Tropheus sp. Sumbu; Population 7.9

Trade name: Tropheus "Sumbu", Golden Rainbow Moorii.

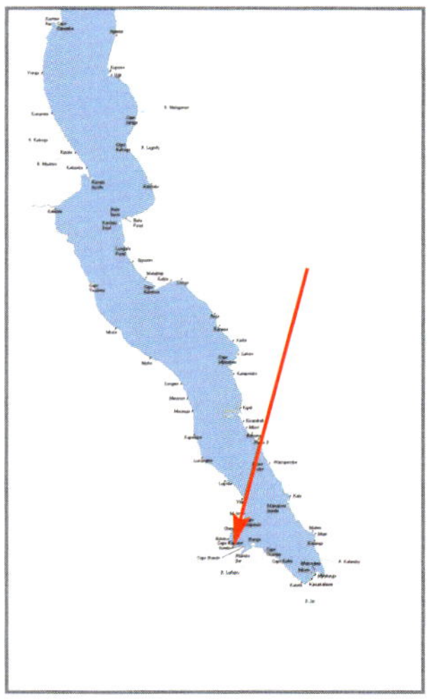

Distribution: The distribution of this local form begins to the south of Cape Kachese and extends right round Sumbu Bay.

Discovered: This colour form was first discovered by German aquarists at the beginning of the 1980s.

First importation: The first importation for the ornamental fish industry took place in the mid 1980s. Unfortunately for unexplained reasons importations of this very attractive local form remain extremely rare. They have found their way to Europe only a few times.

Coloration: Apart from a small greenish area below the dorsal, the body and head alike are a gold-ochre colour. The upper head is only slightly darker. There is a single stripe, sometimes two, the same colour as the body, on the forehead. The iris is bluish with an upper orange-red area. The dorsal is resplendent in a vivid cinnabar red, and the spines are not dark but the colour of the body, ie gold-ochre. The ventrals are dark with red membranes. The anal is orange-red only on the spinous part, becoming ochre-coloured further back. The base of the anal is light violet. Caudal and pectorals are darker, usually greenish-orange.

Remarks: The *Tropheus* in Sumbu Bay are somewhat lighter still than the *Tropheus* from Cape Kachese. Sumbu Bay is relatively shallow and polluted with sediment from the river that enters the bay north of the village of Sumbu. This local form is extremely attractive but little known among aquarists. The individuals imported at the 1980s were solid gold-ochre in colour. Unfortunately such attractive fishes are no longer imported from that site, perhaps because the precise collecting site has been forgotten. The fish was imported at the beginning of the 1990s in a consignment of *Tropheus* sp. "Kachese". It is not so attractively solid-coloured as the Golden Rainbows of those days, but is very similar to them. Because the Sumbu National Park borders the lake, it is possible that crocodiles frequently disrupted the collecting plans of the exporters. The political situation in Zambia is relatively stable and hence it should nevertheless be possible to find and collect these attractive *Tropheus* again.

This adult female is an aberrant form of the Sumbu colour variant. Typical *Tropheus* from Sumbu are solid gold-ochre coloured with no dark dorsal spines.

photo: P. Schupke

Tropheus sp. Cape Nundo; Population 7.10

Trade name: Tropheus "Golden Chisanse".

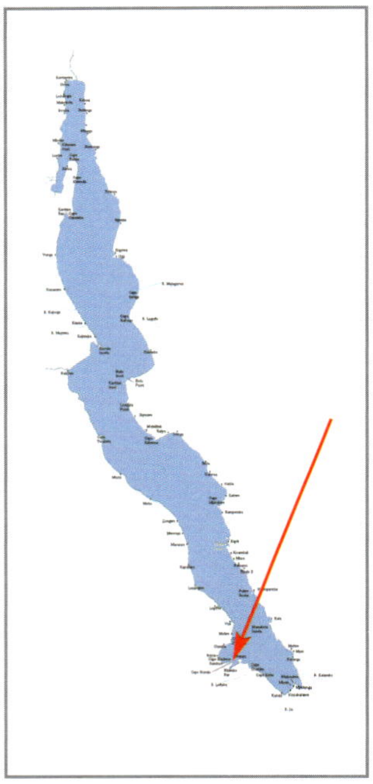

Distribution: The distribution of this local form is given as Cape Nundo, at the western entrance to Nkamba Bay, and extends well into Nkamba Bay itself.

Discovered: This colour form was discovered at the end of the 1980s by ornamental fish collectors.

First importation: This colour form was imported for the ornamental fish industry at the beginning of the 1990s. It has not often found its way into aquarists´ tanks.

Coloration: The coloration is split into two areas: the head and the area above the lateral line are a brownish-greenish colour, while the area below the lateral line is straw yellow. Mouth, throat, and the base of the anal exhibit a light violet colour, also apparent in the stripes on the forehead. The rust-coloured dorsal and anal have a bluish-grey sheen on their posterior parts.
The outer eye ring is yellowish-orange, the inner light bluish. These *Tropheus* can become much darker, depending on mood. Adults normally no longer show any vertical barring, unless threatened by a stronger individual or kept in a bare tank. Females may show vestiges of their stripes even when adult.

Remarks: The *Tropheus* from the Cape Nundo area are somewhat reminiscent of the neighbouring *Tropheus* sp. "Ilangi". Cape Nundo consists of a much-eroded rock-face which includes a large pool that also serves as a hippo "bath-tub". The water there is mostly very clear and would be ideal for snorkelling were it not for the hippo population. If one approaches incautiously, the hippos may see the boat as a rival to be attacked. In addition there are also a number of crocodiles, usually lurking unseen. The scarcity of this colour form in our aquaria is undoubtedly due to these hazards.

Tropheus sp. Chilanga; Population 7.11

Trade name: Tropheus "Chilanga Red", Green-Red Moorii, Tropheus "Nkamba".

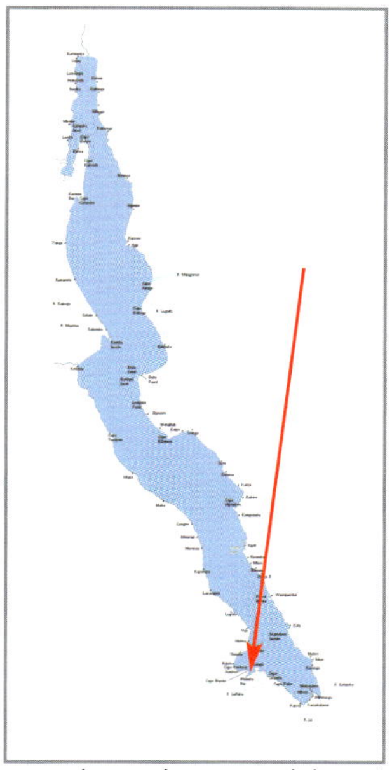

Distribution: The distribution begins at the Nkamba Bay Lodge and extends at least as far as Chilanga Rocks, which lie almost at the end of the eastern bay.

Discovered: This local *Tropheus* form was discovered by German aquarists in the mid 1970s.

First importation: They arrived in the trade in small numbers as early as the end of the 1970s. Later on things went quiet as regards this *Tropheus*, and it was not imported for several years.

Coloration: *Tropheus* sp. "Chilanga" looks rather similar to the *Tropheus* from Chimba. The body is divided into two colour zones: forehead and dorsum have a greenish sheen, while the lower part of the head and the body beneath the lateral line are a vivid red. Depending on mood, these *Tropheus* can become much darker, usually when sexually active. They become very light in stress situations, eg after capture or in newly set-up tanks. The caudal base appears rather greenish yellow. The mouth, pectoral insertion, and base of the anal are a light violet colour. The dorsal, the lower part of the gill-cover, the ventral membranes and the spinous part of the anal are bright red in colour. The iris has a light golden yellow inner ring, while the rest of the eye is red on the upper edge and bluish-silver below.

Remarks: The Green-Red *Tropheus*" is an old friend, which may be regarded as one of the *Tropheus* pioneers. Staeck documented these *Tropheus* for the first time in December 1975. The form was not imported for many years, and it was not until the end of the 1980s that it was again collected for the ornamental fish industry. In 1998 it appeared again in small numbers, along with *Tropheus* sp. "Ilangi", exported to Europe by Chris Blignaut. The Green-Red *Tropheus* was traded as *Tropheus* "Chilanga Red" by another exporter, Toby Veall. Chilanga Rocks is an underwater reef, which lies almost at the end of the eastern Nkamba Bay. H. J. Herrmann mentions a Green-Red *Tropheus*, that he caught near to the Nkamba Bay Lodge. Thereafter this *Tropheus* is found along a long section of coast on the eastern side of Nkamba Bay. Brichard states that the Green-Red *Tropheus* and the Yellow Rainbow *Tropheus* meet at the exit to Nkamba Bay, but that they do not hybridise. At that time, of course, he could not have known that the Yellow *Tropheus* (*T.* sp. "Ilangi", yellow) was a light-coloured variant of the *Tropheus* "Chilanga Red". The rarity of the latter in the aquarium hobby is due to the very large hippo and crocodile population in the southern part of Nkamba Bay.

The whole of Nkamba Bay can be seen in this aerial photograph. Sumbu Bay is in the background, with Cape Nundo in between. The yellow *T.* sp. "Ilangi" inhabits the little bay below right, and the more orange-red *Tropheus* sp. "Chilanga" is found around the bay to the left. **photo**: P. Schupke

Tropheus sp. Ilangi; Population 7.12

Trade name : Tropheus "Ilangi".

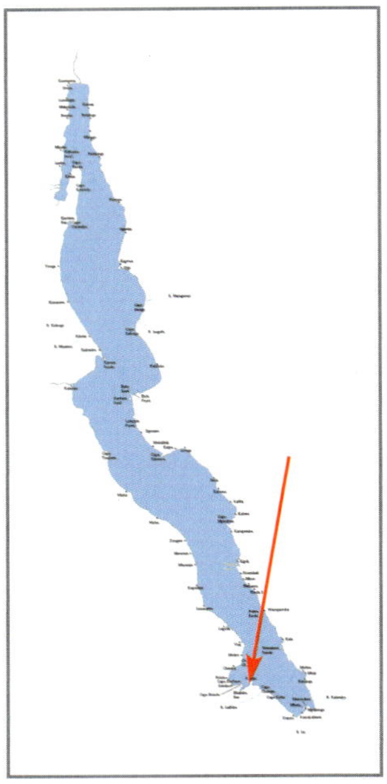

Distribution: The distribution of this local form lies in a small bay a few kilometres from the eastern entrance to Nkamba Bay.

Discovered: I have no details of the discovery of this colourful variant. It was probably first discovered by one of the exporters.

First importation: *Tropheus* sp. "Ilangi" first reached Europe at the beginning of the 1990s, and has subsequently remained permanently on the "want list" of importers.

Coloration: Here too, the base colour consists of two differently coloured zones on the body, but the dividing line between them is not clear-cut and hence less striking. The upper head and the dorsum above the lateral line are greenish. The rest of the head is a brownish-red colour, somewhat overlain with light blue-violet and more distinct on the mouth, chin, and throat. The body below the lateral line is a brilliant yellow, Overlain with a greenish cast above and an orange one below. The dorsal is a vivid light red, and the spines are not delineated with darker colour. Caudal, pectorals, and ventrals are dark, and the ventral membranes are tinted reddish. The bases of the pectorals and anal are light violet. The anal exhibits the red colour of the dorsal, only slightly diluted. The upper part of the eye is yellowish-orange, while the lower part is bluish. During sexual activity the body becomes much darker. Adults generally no longer display any barring. The bars on the forehead characteristic of this lineage are rarely present.

Remarks: The "Ilangi" is collected by two different exporters at two different sites. The specimens collected by Chris Blignaut exhibit a slight reddish tinge, while those from Toby Veall display the yellow colour described above. The reddish form is a so-to-speak aberrant variant of the dark red *Tropheus* from south-east Nkamba Bay. The yellowish type is collected at the entrance to the eastern bay. As already mentioned in the discussion of the lineage on page 75, I obtained fry of two different types from a pair of "Ilangi": three of the nine fry were much lighter (see below), and were more like the "Yellow Rainbow" which is the north-eastern neighbour of *T.* sp. "Ilangi" and extends as far as the Kasaba Bay Lodge. Where the two forms interbreed the resulting brood contains both parent types. Perhaps one of the ancestors of my "Ilangi" was a "Yellow *Tropheus*", nowadays known as *Tropheus* "Nkonde yellow". *T.* sp. "Ngwena Bay", first imported in 2001, may - depending on personal viewpoint - be regarded as a precursor of *T.* sp. "Ilangi".

A juvenile *Tropheus* with greatly lightened coloration, one of the brood of nine fry (as is the fish in the photo at the top of the next page) that exhibited two different colorations. The similarity to the "Yellow Rainbow" cannot be denied.
photo: P. Schupke

EA *Tropheus* sp. "Nangu" or "Yellow Rainbow" for comparison. The similarity to the light-coloured juvenile *Tropheus* sp. "Ilangi" is striking, but does not necessarily indicate a close relationship, as it could equally represent convergent adaptation to a similar habitat.

photo: P. Schupke

Tropheus sp. Linangu; Population 7.13

Trade name: Tropheus "Linangu".

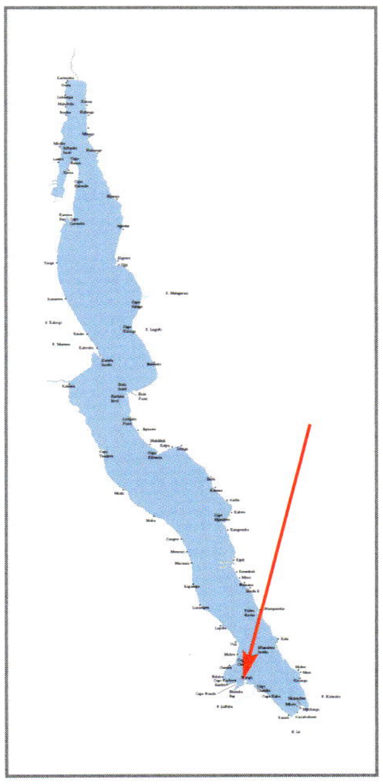

Distribution: The distribution, as currently known, supposedly lies in southern Kala Bay, which is separated from Kasaba Bay only by a few sand dunes.

Discovered: This colour form was probably discovered by aquarists back in the 1980s.

First importation: These attractive fishes did not find their way into our aquaria until the end of the last millennium. As far as I know, these *Tropheus* were first imported by CJ Aquarium.

Coloration: *Tropheus* sp. "Linangu" can be described as a lighter version of *T.* sp. "Ilangi". The brown-red to dark wine-red is striking, as is the very dark, almost black tail. The greenish dorsum adjoins the light-yellow colour of the flank, which extends down to the belly. The caudal peduncle is very light, almost white. Apart from the vivid yellow dorsal fin and the tail, the fins are virtually colourless. Females and non-territorial males have a clearly visible pattern of stripes.

Remarks: *Tropheus* sp. "Linangu" has found few fans since its arrival, perhaps because it doesn´t show its best colours in dealers´ tanks. But anyone who buys this form, especially "old hands" with *Tropheus* who know how best to maintain them, will soon see *T.* sp. "Linangu" in all its glory. In 2001 a very similar colour variant was collected by Thorsten Reuter, a German aquarist who has now made his home in Zambia, and exported for the aquarium trade in Germany. The rather darker yellow of this population extends onto the dorsum and the posterior head, so that the body appears uniform dark yellow. The dark head and tail are striking in this population, which is imported as *T.* sp. "Nkonde Yellow" (Pop. 7.13.1). The intensity of the yellow colour can vary considerably from individual to individual.

Tropheus sp. Nangu; Population 7.14

Trade name: Tropheus "Nangu", Tropheus "Yellow Rainbow".

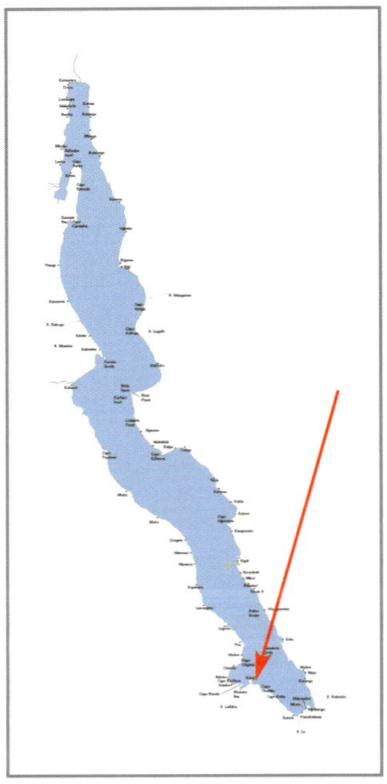

Distribution: The distribution of this colour variant encompasses a large part of the Nangu peninsula. Little is known regarding whether and how much its coloration varies along this long stretch of coastline.

Discovered: This local *Tropheus* form was discovered in 1975 and documented in 1977 by Wolfgang STAECK.

First importation: The first importations took place in 1978.

Coloration: The brown-yellowish base colour lightens to whitish to whitish-yellow on the lower half of the body. The dark brown-reddish head has a greenish area on the forehead. Apart from its yellowish upper part the iris is light silvery. Depending on mood, narrow, whitish-yellow vertical bars are displayed, becoming lighter towards the belly. Even narrower, but much fragmented, stripes are visible in the intervals. The yellowish grey-blue dorsal has a narrow marginal band. The spinous part of the anal has an orange-red dot on the membrane between each pair of spines, and a few small egg-spots can be seen on the posterior part. Caudal, pectorals, and ventrals are generally dark. The pectoral base is pink. At the Nangu peninsula the light blue dots on the head are absent and the colours are not as vivid as in individuals from Cape Kabeyeye.

Remarks: The Yellow Rainbow *Tropheus* is much more attractive than might, unfortunately, be thought from (too) many of the photos in the literature. Its coloration is very much dependent on how it feels. Poor water quality has a negative effect on the intensity of the colours. Some years ago I saw a picture of a *Tropheus* in an English aquarium magazine that immediately reminded me of the Yellow Rainbow, but the colours were much more intense. The yellow was more of a yellow-orange, and the most striking feature was a yellow spot on the central flank. The photo also showed a collecting team on Lake Tanganyika. However, when I asked an importer which exporter´s collecting team was shown in the photo, no additional information was forthcoming. Many years later this *Tropheus* appeared in the trade as *T.* "Linangu". The relatively long coastline around Cape Nangu is not a good place to visit, as it is the haunt of numerous crocodiles. I am sure there must be colour gradations within this 25 kilometre long stretch of coast, just as colour variations within a population are known elsewhere. The colour variants *T.* sp. "Linangu" and *T.* sp. "Nangu" are, so to speak, lighter forms of *T.* sp. "Ilangi", according to the latest DNA research. This signifies that these two colour forms do not belong to lineage 8, as one might assume at first glance, but to lineage 7.

Tropheus der Linie 8

The *Tropheus* of this lineage arrived in our aquaria at a very early stage. To start with there were only a few colour forms, but, because of the pertinacity of fans of these fishes, new stretches of coastline were forever explored. In those days any unknown stretch of shore meant a new colour variant of these fascinating fishes. And even now the coastline is far from being thoroughly explored. Because of the enormous size of Lake Tanganyika and the highly variable structure of its coastline (ie the underwater landscape), it is unlikely that even today all the colour variants have been discovered. Thus there must almost certainly still be a few populations with very limited distributions still waiting to be discovered. The distribution of this lineage and its boundaries are, however, largely well known today. We know of at least 21 well differentiated populations in this lineage, known by the popular name of "Rainbow *Tropheus*". In the south-western part of the lake, lineage 8 may start just to the south of Kasaba Bay. Its distribution initially continues further south and then crosses the Tanzanian border. The area north of the border remained unexplored for a long time. Not until 1983 did aquarists based at Ndole Bay make their way as far as the Malasa Islands. The *Tropheus* there showed no major differences from those further south, and it was thus not particularly difficult to assign them to the Rainbow *Tropheus*. Not until the beginning of the 1990s did the coast begin to be systematically explored progressively from the south. The coastline was now explored from Kasanga to Karema, just south of Ikola. In the process the northern limit of the distribution of the Rainbow *Tropheus* was also discovered: as far as we know at present it lies at Kisambali, a few kilometres south of Kipili, where its distribution supposedly overlaps that of another lineage, lineage 9, for about 10 kilometres. The coloration of the individual local forms along this approximately 300 kilometres of coastline is correspondingly variable. However, they all have a number of characters in common that indicate that they belong to a single lineage. A common feature of all the southern Rainbow *Tropheus* is a more or less clearly visible elongate light patch on the central lower flank, which may sometimes extend upwards almost to the dorsal. Depending on the locality of the form in question, this patch may be whitish, yellow-white, yellow, yellow-orange, or orange. The colour of the dorsal is just as variable - it can be yellow, red, blue, violet, or any intermediate shade. Then there are breathtaking combinations, incomparable in their intensity of coloration. The head is adorned with numerous tiny dots; this feature is seen mainly in the Zambian and Congolese colour forms. The dorsum is often greenish, or somewhat brownish in the forms from northern Tanzania. The caudal is dark and only slightly indented. With two exceptions, all forms have 6 anal spines. The stripe pattern of this group of *Tropheus* is very characteristic: the stripes are narrow, and in juveniles there are additional, very narrow, often incomplete, intermediate stripes. In adults the stripes are always narrower and less complete, and intermediate stripes can also sometimes be seen. For many years a huge variety of colour forms from Zambia have been offered for sale. *Tropheus* such as the "Red Rainbow", the "Orange Flank", and the "Lemon Flank" are real export hits. Less popular colour forms are not always available, but are invariably imported again after a long interval. Because these fishes have been readily available for many years, it might be assumed that imports would have slowly diminished, but this has not been the case. There is no shortage of tank-breds, but these are not as regularly purchased as the so-called "wild-caughts". Larger tank-breds often have a far less convex upper head profile and a less cylindrical body not seen in individuals from the lake, where they must "work" very hard each day and the feeding musculature of the head is correspondingly well-developed.

According to the latest DNA research the *Tropheus* of lineage 8 derive from the splitting of several lineages. In the course of time, as the lake level has fluctuated, their genes have mingled to create a new lineage with a number of discrete characters in common.

Lineage 8 contains the populations numbered 8.0 - 8.22.

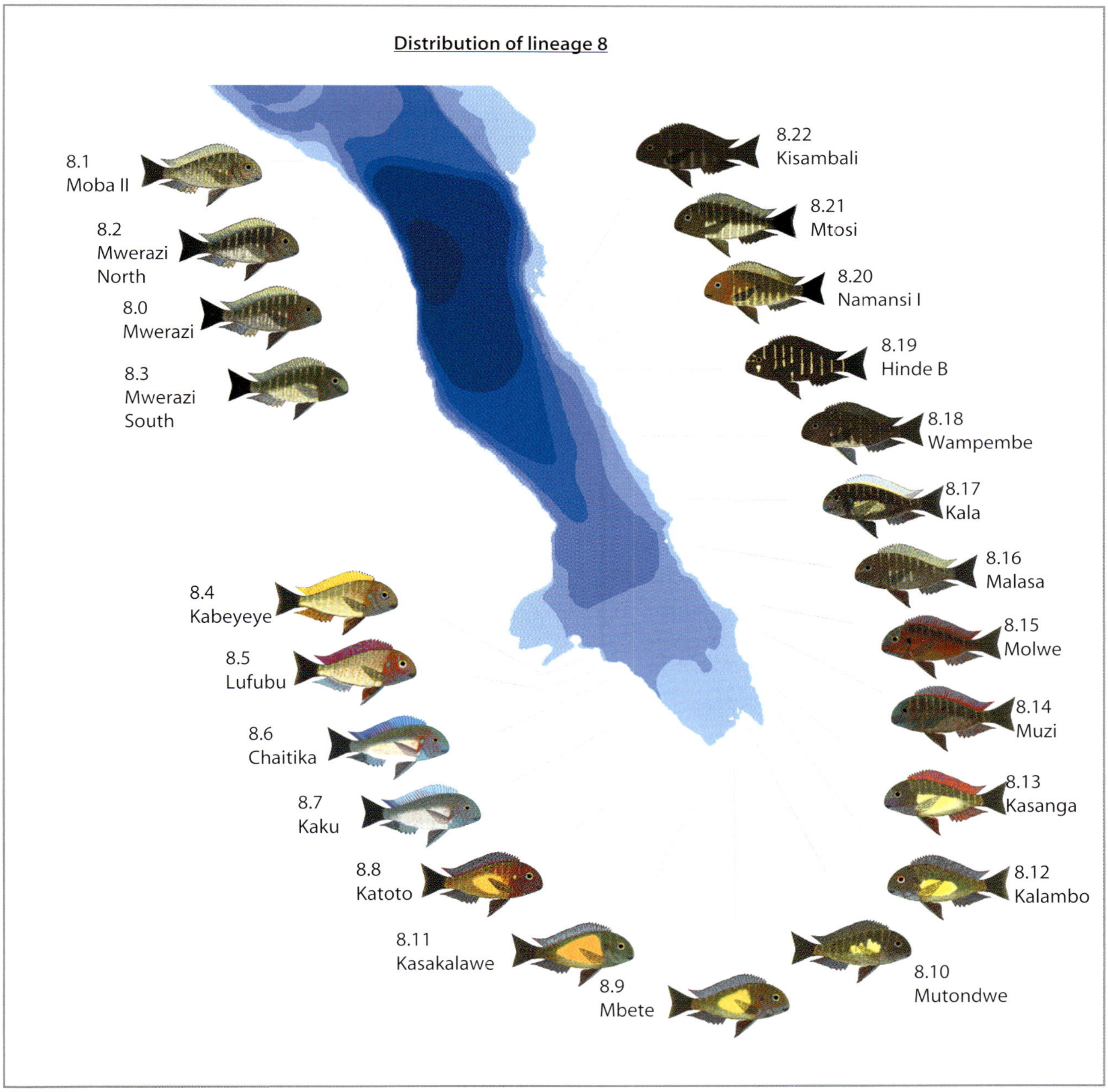

Tropheus sp. Moba II (Murago); Population 8.1

Trade name: Tropheus "Murago".

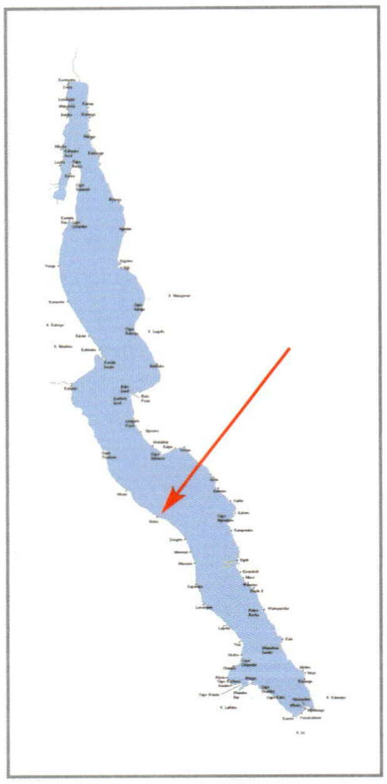

Distribution: This colour form was found at the harbour wall at Moba. It is not known whether this location is a natural site for this population. See also Remarks.

Discovered: This *Tropheus* population was discovered by Pierre BRICHARD and his collecting team at the beginning of the 1980s.

First importation: At the end of the 1980s, *Tropheus* of this form, pond-bred by the lake by BRICHARD´s company, were available in the German aquarium trade.

Coloration: The base colour is grey-green, with yellowish-white vertical bars that change to yellow and become wider on the lighter, almost white, lower part of the body. The strikingly whitish-yellow dotted head is more brown on its upper half and pink on the lower. The upper lip is light blue. The dorsal is yellow in females, but more of a light blue in males; in both sexes it is light yellow along its entire base. The caudal, like the ventrals, is a dark grey colour, and has a narrow black marginal band. The spinous anal is light blue, while the soft part is more transparent. The lappets of the spinous part are an orange shade. The iris is silvery apart from an upper yellowish area. Juveniles too have dots on the head, but these are not as striking as in adult specimens. They resemble the "Yellow Rainbow" (Pop. 8.4) in the distribution of colour on body, head, and fins.

Remarks: For commercial reasons Pierre BRICHARD kept the source of this *Tropheus* a great secret. Other collecting teams that gradually explored the coastline of the former Zaire found the "Murago" in several places. For example, a *Tropheus* identical with BRICHARD´s "Murago" was found on the harbour wall at Moba. Another *Tropheus* with a heavily dotted head is considered to be a subpopulation of *Tropheus* sp. "Mwerazi North" (Pop. 8.2). Assuming that the man original collecting site at Moba is correct, then, the distribution of the "Rainbows" on the west coast is extended considerably. The Muragos at Moba may be sympatric with the fork-tailed *Tropheus* of lineage 4. Like all other local forms of this lineage in the Congo, the Muragos have only 5 anal spines, as opposed to the six anal spines of the Rainbows of lineage 8 in Zambia and Tanzania. The origin of these northern Rainbows remains unclear, as no research material is available. At the present time no collecting is taking place along that stretch of coastline.

In this close-up head study not only the dots, but also the red-tipped teeth in the mouth, are readily visible. **photo**: P. Schupke

Tropheus sp. Mwerazi North; Population 8.2

Trade name: Tropheus "Mwerazi".

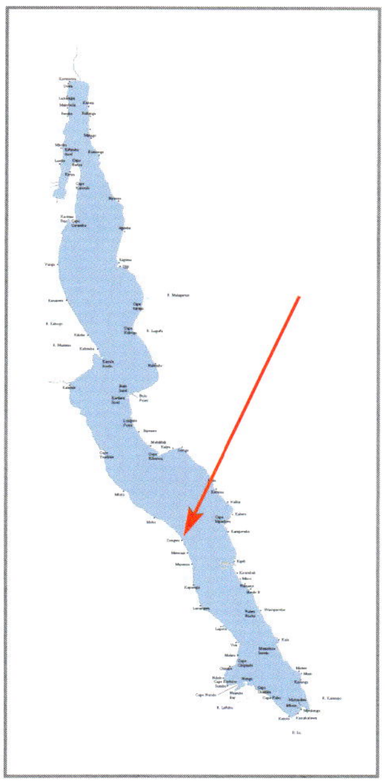

Distribution: The distribution of this local form begins a few kilometres north of the village of Zongwe and supposedly extends to Zongwe itself.

Discovered: This *Tropheus* colour variant was discovered by Max Poll during his major hydrobiological expedition in 1946-47, and documented by him in 1956.

First importation: This *Tropheus* was first imported for the German ornamental fish industry in 1994. Further imports have been rare because of the political situation in the Congo.

Coloration: The grey-green base colour of the body tends to green-grey on the head. The lower half of the head is a delicate shade of pink, which continues as far as the breast. The entire head, as well as the breast and throat regions, is covered with whitish dots of variable size. The area from the snout to the upper head is rather greenish. The body exhibits yellowish-white vertical bars which do not extend onto the belly. On the central flank there are very narrow, barely visible, stripes in the intervals. The stripes may extend into the lower part of the dorsal. There is a small brown-red spot above and in front of the pectoral, between it and the upper part of the operculum. The caudal, like the rays of the otherwise transparent pectorals, is very dark, almost black. The anal and ventrals are a grey-blue colour. The anterior two thirds of the dorsal are a vivid yellow, which may shimmer gold-greenish or gold-bronze depending on the light. The posterior part is uniform grey-blue. The belly region is very light, sometimes even white. These fishes have five anal spines.

Remarks: Not much is known regarding the distribution of these north-western "Rainbow *Tropheus*". The approximately 50 kilometre stretch of coast where these *Tropheus* were found is home to several colour variants, the limits of whose distribution remains unknown. Likewise unknown are the limits of the *Tropheus* of lineage 6, which may occur within the range of these Rainbows (it is impossible to be more precise because of the lack of accurate information), or perhaps their distributions merely overlap. In the south the *Tropheus* from Kapampa adjoin the southernmost limit of the range. The discovery of the Murago at Moba means that *Tropheus* sp. "Zongwe", which belongs to lineage 6, occurs in the middle of the range of these Rainbow *Tropheus*. According to information from one importer, population 8.2 contains a sub-population (catalogued here as 8.0, no text) that resembles the Murago.

The difference between the two populations of the north-western Rainbow *Tropheus* can clearly be seen by comparing these two fishes. The lower fish belongs to population 8.2, supposedly found in the general area of Mwerazi, while the upper belongs to subpopulation 8.0.
photo: P. Schupke

Tropheus sp. Mwerazi-Süd; Population 8.3; *Tropheus* sp. Mwerazi; Population 8.0;

Trade name: Tropheus "Mwerazi".

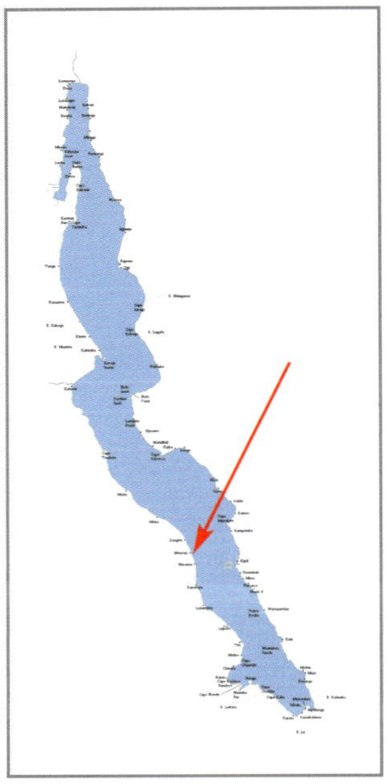

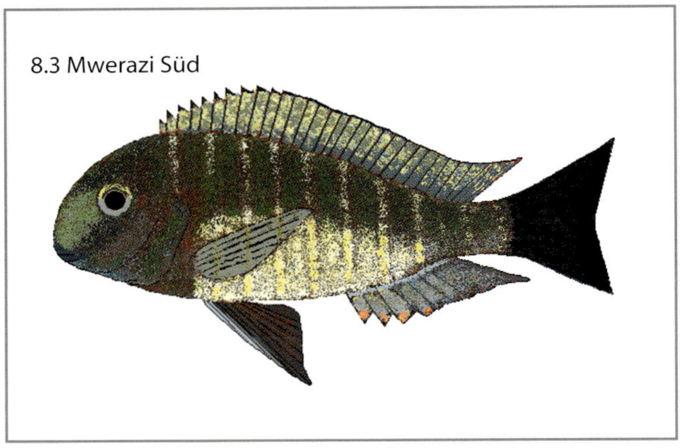

Distribution: According to Pierre Brichard, the distribution of these local forms lies to the north and south of the village of Mwerazi on the southern section the Congo coast.

Discovered: The *Tropheus* of Mwerazi (8.3) may have been discovered by Max Poll during his expedition in 1946-47.

First importation: These *Tropheus* were first imported to Germany in 1994 by CJ Aquarium.

Coloration: The region from the upper lip, over the forehead, and along the entire dorsum is emerald green. The gill-cover has a pink to greenish tint, which may turn to brown-red on the upper posterior operculum and then extend behind the gills to the pectoral insertion. The yellowish-white vertical barring, which includes intermediate stripes, is rather bolder on the lower, whitish half of the body and creates the impression of a yellow spot. Only when the fish is excited does the yellow-white underside appear as a striking patch, as the rest of the body becomes much darker, much as in sexually active *Tropheus* sp. "Kalambo". The head (apart from the greenish forehead) and breast now become more of a wine-red. The grey-blue anal is bordered at its base by a more vivid light blue-grey. The caudal is very dark, almost black. The leading edges of the ventrals are blue-grey, while the membranes appear dark wine-red. The smoky dorsal has a rather gold-yellow sheen and a narrow black marginal band. The upper part of the iris is yellow, and more silvery below.

Remarks: This local form includes a sub-population (here labelled 8.0) which is supposedly very similar to the Murago (8.1). In the southern part of its range the Mwerazi *Tropheus* is said to adjoin the distribution of *Tropheus* sp. "Kapampa". It would be interesting to know if the two lineages (6 and 8) are actually in contact or whether a natural barrier exists between them. Unfortunately these *Tropheus* have been imported only once, which is rather difficult to understand as they are attractively coloured. Some of the mood-related colour patterns of this local form exhibit a considerable similarity to those of the southern Rainbow *Tropheus*.

Fishes are not the be-all and end-all. The water cobra (*Boulengerina annulata*) is another beautiful inhabitant of Lake Tanganyika.
photo: F. Schäfer

Tropheus sp. Kabeyeye; Population 8.4

Trade name: Yellow Rainbow, Tropheus "Nangu", Tropheus "Kabeyeye".

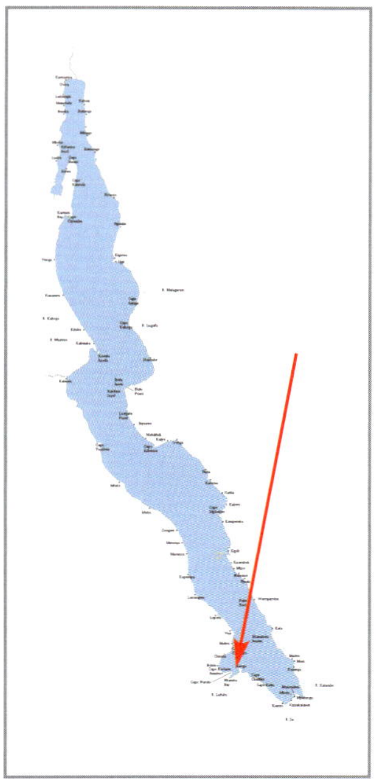

Distribution: The distribution extends to the north of the mouth of the Lufubu, all around the Nangu peninsula to south of Kasaba Bay Lodge, which lies on the far side of Kasaba Bay.

Discovered: This local *Tropheus* form was discovered by Wolfgang STAECK in 1975 and documented by him in 1977.

First importation: The first importations took place in 1978.

Coloration: The grey-greenish base colour becomes a much lighter yellowish white on the lower half of the body. The head is finely dotted with light blue on a reddish-brown background; the latter colour is amplified on the anterior upper part of the body in the form of a reticulated pattern that may sometimes extend rather further back. The red zone is lighter behind the eye, extending across the upper operculum to the pectoral insertion and somewhat further down. Depending on mood, narrow, whitish-yellow vertical bars may be seen; these are light yellow lower down. Even finer stripes are visible in the intervals, but they are much fragmented. The yellowish to yellowish-grey dorsal has a narrow marginal band. The spinous part of the anal has an orange spot on each of the membranes, and a few small egg-spots may be visible on the posterior part of the fin. Caudal and pectorals are generally dark. The ventrals have a rather yellowish-grey sheen but are otherwise transparent. The light blue dots on the head are apparently seen only in individuals collected between Cape Kabeyeye and the Lufubu river.

Remarks: This population may be more closely related to lineage 7, but further DNA studies are required before any conclusive verdict on this fish can be given.

Tropheus sp. Lufubu; Population 8.5

Trade name: Tropheus "Lufubu".

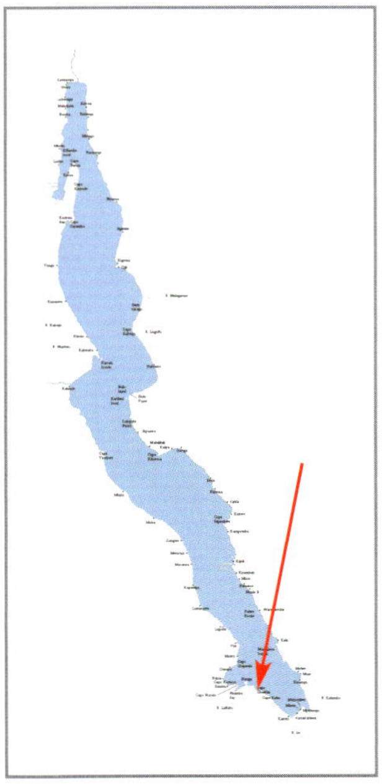

Distribution: The distribution begins to the east of the mouth of the Lufubu and extends only a few kilometres further east.

Discovered: Pierre BRICHARD mentions this colour form in his book, "Cichlids and all other fishes of Lake Tanganyika", and probably discovered it during his collecting trips to Zambia at the beginning of the 1980s.

First importation: As far as I know, this *Tropheus* was first imported for the ornamental fish industry at the end of the 1980s.

Coloration: The lower half of the wine-red head is overlain with a light violet sheen. There are light blue dots not only on the head, but also on the throat and breast region. A narrow greenish stripe runs upwards from the upper lip and along the dorsum to the caudal peduncle. The body is beige-whitish with a wine-red reticulation which becomes fainter posteriorly. The dorsal is mainly red-violet with a few blue inclusions and a narrow dark marginal band. The upper and lower edges of the caudal exhibit reddish tints. The rays of the upper half of each pectoral and the soft parts of the ventrals are also red-violet. There are light blue areas on the rays of the pectorals and anal, which also has red-violet zones. The posterior edge of the anal exhibits a few small, barely discernible egg-spots. The silvery iris has a narrow orange-yellow edging and a spot of the same colour in its upper part.

Remarks: This local form is one of the most colourful colour variants in the lake. Such intensely coloured individuals are found for only a few kilometres east of the Lufubu river, and by Cape Chaitika differently coloured specimens are collected. For several years the Lufubu *Tropheus* was known only to absolute insiders, and other *Tropheus* fans knew only the name. These fishes are very difficult to catch in their native biotope, as they live among large blocks of rock. In the mid 1990s I saw an advert offering *Tropheus* "Lufubu" for sale, and was rather surprised that it was necessary to advertise *Tropheus* belonging to one of the most coveted colour variants. But when I saw these fishes it was immediately clear why no-one wanted them. They were sooty black, exhibiting no colour at all. It turned out that at that time the water in Würzburg came out of the tap with a nitrate content of about 80 mg/l. II bought these fishes. After only a few days they began to become lighter, and after a few weeks the exhibited a very attractive coloration (see photos).

The *Tropheus* showed no colour in water with a high nitrate content.

After a few weeks in good quality water that had changed completely.

photos: P. Schupke

Tropheus sp. Chaitika; Population 8.6

Trade name: Blue Rainbow Moorii, Tropheus "Chaitika".

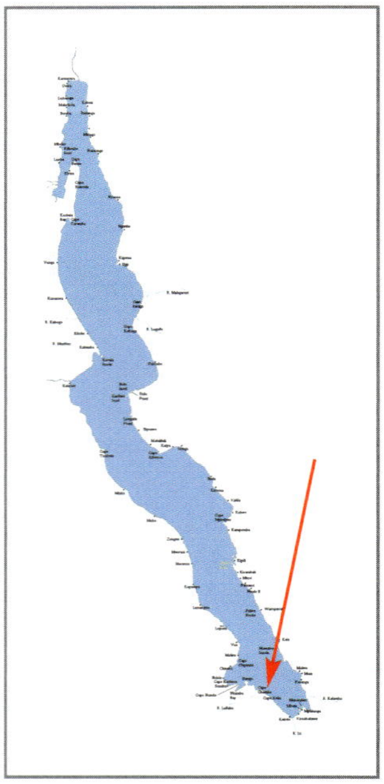

Distribution: The distribution begins round about Cape Chaitika and extends for a few kilometres further south.

Discovered: This local form was discovered by German aquarists in 1975 and documented the same year by Scheuermann and Staeck.

First importation: These *Tropheus* were imported in small quantities for the trade as long ago as 1975.

Coloration: Forehead, upper head, and dorsum are jade green. The central flank is decorated with a light beige-coloured patch. The belly is white. There is a red area beginning behind the eye and continuing onto the body as a red reticulation which, however, loses much of its intensity posteriorly. The upper head is sprinkled with small green spots and lines, changing to light blue dots lower down. The area above the upper lip is blue-turquoise, while throat and chin are light turquoise. The largely vivid blue dorsal has red on the rays of its posterior part. The light blue anal is decorated with a number of red dots. The ventrals are bluish anteriorly, otherwise dark. Pectorals and caudal are very dark, but with red tints faintly discernible. The silvery iris has a yellow-orange outer ring, more extensive above. Sexually active males become very dark apart from the light patch on the belly.

Remarks: This *Tropheus* is one of the most beautiful colour variants. Its coloration remains very consistent in the Cape Chaitika area, but only a few kilometres away it becomes noticeably paler and the brilliance of the colours is lost. The *Tropheus* from Cape Chaitika loses its colour very quickly if maintenance is less than optimal, becoming paler or assuming a darker shade. The beautiful colours that cause us to marvel in the aquarium are barely recognisable underwater in the lake. The red shades are difficult to perceive in as little as one or two metres of depth, and suspended particles finish the job. The rocky littoral at Cape Chaitika is densely populated with *Tropheus*, which are generally found about two metres apart. I was unable to detect any quarrelling; the fishes were so busy feeding that there was not much time for fighting. It appeared that only males occupied territories while females were more inclined to wander around. This colour form was one of the first *Tropheus* to be offered in the trade at the end of the 1970s. Later on the Cape Chaitika *Tropheus* regularly appeared on dealer´s lists and its popularity remains high. The colours are best appreciated in well-lit aquaria.

Underwater scene at Cape Chaitika, with a sexually active male in the centre of the picture. **photo**: P. Schupke

Tropheus sp. Kaku; Population 8.7

Trade name: Tropheus "Kaku".

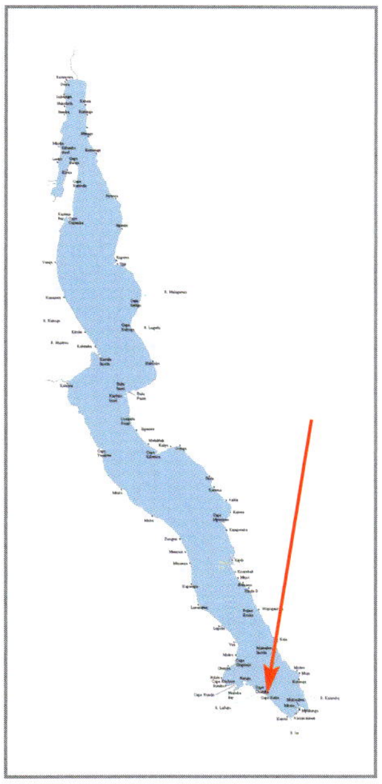

Distribution: The distribution of this colour form begins a few kilometres south of Cape Chaitika and ends a lot further south.

Discovered: This *Tropheus* was discovered at the end of the 1980s during exploration of the coast.

First importation: *Tropheus* sp. "Kaku" was collected in small quantities for the ornamental fish industry at the end of the 1980s. Subsequent imports have been seldom, and to the best of my knowledge no stocks have been imported in recent years.

Coloration: Forehead, upper head, and dorsum have a light greenish cast. The reddish-brown of the head is almost devoid of bluish dots. The head colour continues along the body in a reticulated pattern, becoming fainter posteriorly. The beige-whitish lower body normally no longer exhibits vertical bars, but, depending on mood, for example in stress situations, these may reappear. Dorsal and anal are bluish anteriorly, becoming more grey-blue posteriorly. The anterior dorsal spines may exhibit reddish tints. Caudal and pectorals are dark, contrasting with the body. There is a small orange-reddish area behind the gills and around the pectoral insertion. The iris appears yellowish in its upper half. The mouth is dark, and the chin a lighter, shade of blue.

Remarks: *Tropheus* sp. "Kaku" is, in terms of colour, a much-diluted form of the *Tropheus* from Cape Chaitika. It would be very interesting to discover why there is such a dramatic change in colour within a limited stretch of coast. One possible reason is the presence of plankton (which clouds the water). At Cape Chaitika, however, the water is relatively clear at most times of the year and 10 metres or more of visibility is nothing uncommon. A few kilometres further south, where colour intensity is already noticeably reduced, the situation is little different. Likewise the water is relatively clear to the north towards the Lufubu. In my opinion, predation pressure from above by piscivorous birds is likewise irrelevant as in clear water the intense colours are relatively unimportant as regards camouflage, and dark fishes are more noticeable than colourful ones. This is undoubtedly an area of research that will provide answers regarding a number of interesting aspects of intraspecific communication and evolution.

Tropheus sp. Katoto; Population 8.8

Trade name: Tropheus "Katoto".

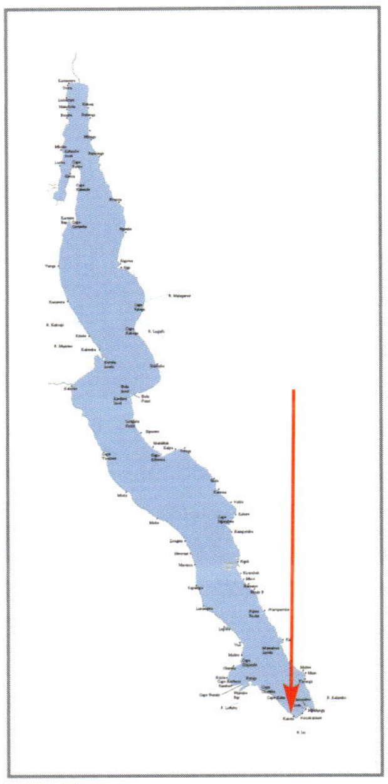

Distribution: The distribution of this colour form is restricted to a few kilometres to the east of Mbete in Zambia.

Discovered: *Tropheus* sp. "Katoto" was discovered by H. W. DIECKHOFF in 1987.

First importation: This *Tropheus* did not appear in the trade until the beginning of the 1990s, and is rarely imported.

Coloration: EA yellowish-orange patch, without clear-cut edges, extends over almost the entire flank, in some individuals even extending to the base of the dorsal. The upper part of the body is a greenish colour, overlain with a faint, light wine-red reticulation. The entire head, including throat, breast, and the area between gills and pectorals, is likewise light wine-red. The wine-red area is covered with very striking blue-grey dots. The iris has an orange edging. The dorsal is light blue with a wine-red base and rays, with the amount of wine-red increasing posteriorly. Depending on mood, adults may still exhibit vertical barring. The bluish anal and pectorals are also wine-red, but the colour is rather more diluted. The dark caudal has a perceptible faint wine-red cast. The underside is very light, almost white.

Remarks: This very attractively coloured local form was discovered relatively late on. Short stretches of coastline, sometimes ignored during exploration because, above water, they appeared unsuitable for *Tropheus*, may even today harbour as yet undiscovered colour variants. This was the case with *Tropheus* sp. "Katoto". The first specimens did not reach us until the beginning of the 1990s. There are individuals that have heavily speckled heads and in some ways resemble the *Tropheus* found at Mwerazi in the Congo. But the *Tropheus* near the mouth of the Lufubu are also similar, although their large dots are not as striking because they are somewhat darker. DNA studies of the southern Rainbows indicate that these populations unite the genes of different lineages. *Tropheus* sp. "Katoto" is so-to-speak a mixture of the two southern sub-lineages based on *T.* sp. "Chaitika" and *T.* sp. "Wampembe".

Tropheus moorii BOULENGER, 1898; Population 8.9, Mbete

Trade name: Tropheus "Mbita", Lemon Spot Moorii, Tropheus "Yellow Spot".

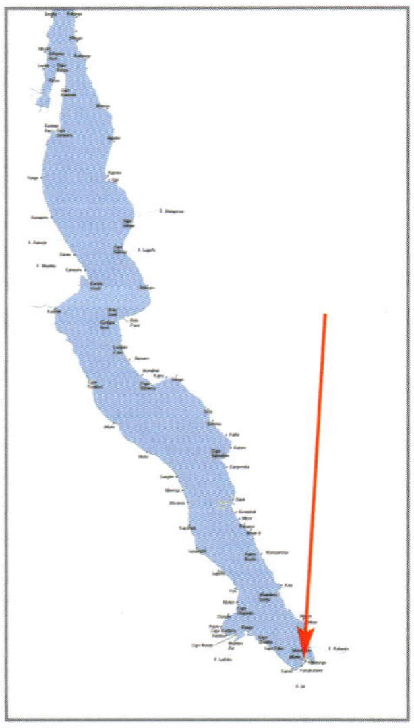

Distribution: The distribution encompasses the general vicinity of and includes Mbity Rocks, also known as Kumbula Island.

Discovered: This *Tropheus* formed the basis for the description of *Tropheus moorii*. MOORE collected it in 1895-96. It was rediscovered for the aquarium hobby by Wolfgang STAECK in spring 1975.

First importation: This local form was being imported by the beginning of the 1980s.

Coloration: The base colour consists of an olive-green, which, in aroused individuals, becomes very dark, virtually black. There is a yellow patch, somewhat overlain with orange, on the central flank, contrasting dramatically with the darkened body. The dorsum appears more greenish, more evidently so near the yellow patch. As in the preceding colour variant, the head is reddish-brown, appearing somewhat overlain with pink-bluish anteriorly. There are only odd light-blue dots. The eye is edged with yellow-orange, with more colour in the upper part. The grey-blue dorsal is barely perceptibly wine-red at the base of the spines. The anal is rather darker with a light grey-blue zone at its base. The ventrals are dark, with light blue leading edges.

Remarks: The specimen which Moore collected at Kinyamkolo in the south of the lake, and which BOULENGER used as the basis for the description of *Tropheus moorii,* was probably from what is now Mpulungu. The *Tropheus* from the general area of Mpulungu also include the *Tropheus* "Orange Spot" from Kasakalawe (8.11), only a few kilometres further west but does not attain the same intensity of colour. In neutral mood, *Tropheus moorii* appear rather greenish. These fishes vary very little within their distribution. In some places the patch is yellower and the head appears less red-brown. Because the other local forms are more attractively coloured, this variant is rarely imported and correspondingly uncommon in our aquaria.

Adult males of the Mbete population are very similar to the *Tropheus* "Orange Spot".
photo: P. Schupke

Tropheus sp. Mutondwe; Population 8.10

Trade name: Tropheus "Mutondwe".

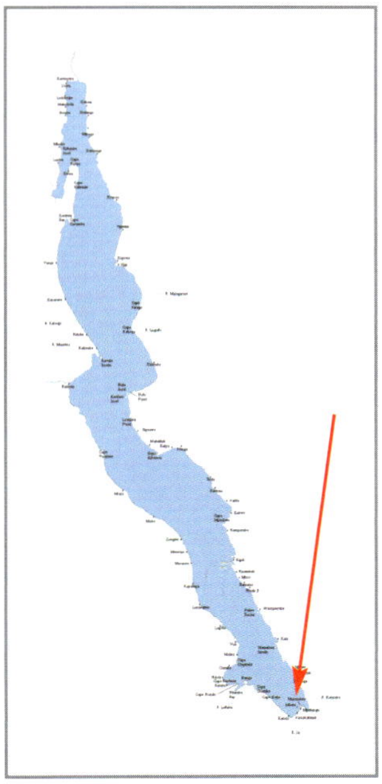

Distribution: This local form is found only at the island of Mutondwe, still also known as Crocodile Island.

Discovered: This local form was discovered by aquarists at the beginning of the 1980s.

First importation: As far as I know, this form has been imported only a few times for the ornamental fish industry, in the 1980s. It is no longer imported today.

Coloration: The base colour consists of a brown-olive shade, which can become much darker depending on mood. As in the colour variants from the mainland, the dorsum exhibits green tints. The head is likewise red-brown, although in a more dilute form and without dots. The dorsal is blue-grey. The yellowish lateral patch is more of a straw-yellow and smaller than in mainland individuals; it varies somewhat in shape and may be split into two or three smaller and more elongate fragments. Mouth and chin, as well as the base of the anal, are blue-grey in colour. The iris is silvery, and the upper part of the eye yellow. The caudal is very dark, as in other populations in this area.

Remarks: The *Tropheus* from Mutondwe Island has basically the same coloration as its relatives from the mainland, Again, a clue as to which form(s) those are would be helpful! MB but in much diluted form. The question of the biological significance of this variable colour intensity cannot be answered here. Perhaps the environment in the rocky littoral plays a part. Unfortunately the modest coloration of the Mutondwe *Tropheus* has won it few fans. At the end of the 1970s I bought a *Tropheus* which was very similar to the Mutondwe *Tropheus*, except for the colour of the dorsal fin, which was red-brown. This coloration was more like that of the *Tropheus* of lineage 7 such as are found in Nkamba Bay. But I was able to ascertain at the time that my fish came from the south. I have never seen this "Lemon Spot" colour variant since. The first exporter to collect fishes down there was Adrian Karr, who left the business many years ago. Perhaps it was a small population whose location was known to him alone.

Tropheus sp. Kasakalawe; Population 8.11

Trade name: Tropheus "Orange Spot", Tropheus "Sonsitt".

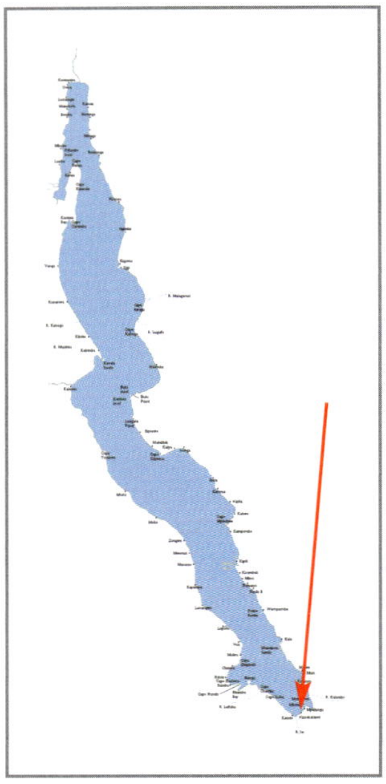

Distribution: The distribution is said to be in the vicinity of the village of Kasakalawe. The stretch of shore inhabited by this attractive colour variant is very limited.

Discovered: This local form was discovered by Wolfgang STAECK in 1975 and documented in the same year.

First importation: This attractive colour form was imported for the ornamental fish industry as long ago as the beginning of the 1980s.

Coloration: The base colour is mood-dependent, as in many *Tropheus*. The most striking feature is the large, centrally positioned, orange patch on the flank. In well-coloured specimens this begins just behind the gills and does not end, horizontally, until after the start of the anal. Vertically it begins immediately beneath the dorsal and ends down on the belly. The head, which again can be described as wine-red, is, like the rest of the body, overlain with a green hue. The dorsal is green-grey with a wine-red base, and posteriorly this colour extends rather further onto the fin. The caudal and part of the caudal peduncle are almost black. The anal has red dots here and there on the spinous part, but is otherwise smoky grey-black. The black pectorals are red at the base. The eye has a narrow orange ring and is the same colour in its upper part. Aroused individuals can become much darker, and in stress situations the flank patch can become sooty.

Remarks: This local form is one of the most colourful colour variants known. It is a staple of ornamental fish dealers, and is imported and bred in large numbers to the present day. The coloured patch is exceptionally large in the fish in the illustration - normally it is rather smaller in area. Tank-breds attain a body form approximating to that of wild-caughts only under optimal water conditions and with plenty of exercise during feeding. If the jaw musculature is under-utilised then the fishes will not develop the high, convex forehead seen in wild fishes. Moreover the body form of wild-caught specimens is not necessarily permanent - if they are unable to pursue their natural feeding habits in captivity the musculature of the head and dorsum will slowly atrophy.

This elderly wild-caught female shows signs of unsuitable maintenance. The musculature has atrophied. The body depth has decreased. The green sheen on the head has also virtually disappeared. **photo**: P. Schupke

Tropheus sp. Kalambo; Population 8.12

Trade name: Tropheus "Kalambo", Lemon Spot Tropheus.

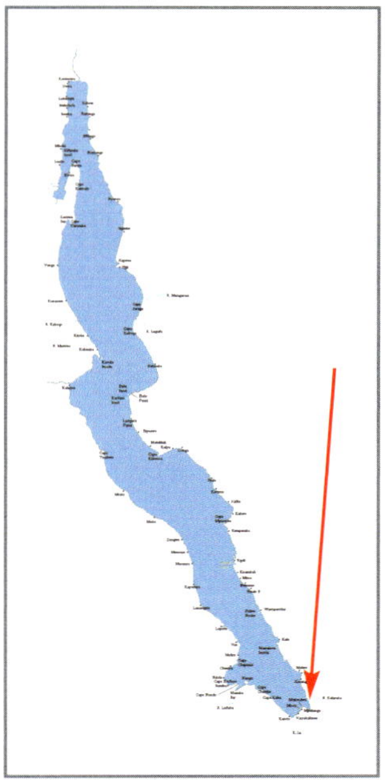

Distribution: The distribution begins to the south of the Kalambo river and ends to the north of Chituta Bay.

Discovered: *Tropheus* sp. "Kalambo" was discovered by Wolfgang STAECK in 1975.

First importation: This local form was imported for the ornamental fish industry at the beginning of the 1980s. These *Tropheus* are commonly seen in the trade.

Coloration: The greenish grey-brown base colour can become much darker depending on mood. This happens mainly during sexual activity or threat behaviour in dominant individuals. The brown-red head exhibits only a few dots. Mouth, throat, breast, and anal base are overlain with light blue. Snout, upper head, and dorsum exhibit a greenish sheen. The horizontally elongate, lemon-yellow patch extends downwards from the centre-line of the flank and contrasts boldly with the olive-black body. The yellow colour is cut off cleanly at this imaginary line. Adults often display traces of their vertical barring. The eye has a narrow edging what colour. also eye should go last in text. MB and its upper half is completely red with silvery flecks. The grey-brownish dorsal has numerous light blue streaks and dots between the rays. The anal is basically grey-blue, with red to orange-red dots that are larger on the spinous part of the fin, smaller on the soft part. Caudal and pectorals are dark.

Remarks: The Lemon Spot *Tropheus* has been known to aquarists for a long time. Customers are always taken with its bright lemon-yellow spot and in particular the red eyes. Even so, it is not imported in such large quantities as its neighbour, the Orange Spot Moorii. Now and then much lighter individuals of *Tropheus* sp. "Kalambo" occur in the lake, and these have captured the attention of exporters and some aquarists. The phenomenon known as xanthism occurs in a number of species in the lake, though it is extremely rarely seen. In such cases the cells that control the colour of the fish lack melanin, the pigment responsible for the darkening of the body. However, dark spots are sometimes still present, and pure orange or yellow individuals are very rare. There is a breeding program to try and breed the "Golden Kalambo" true. Apparently there are aquarists who are prepared to dig deep into their pockets for these "golden calves". Breeding them has proven far from simple as their genetic inheritance is as yet poorly understood.

Tropheus sp. Kasanga; Population 8.13

Trade name: Tropheus "Kasanga", Red Rainbow Moorii.

Distribution: The distribution begins to the north of the border river Kalambo and ends at a long sandy bay shortly before Kasanga.

Discovered: No details available.

First importation: The first *Tropheus* of this colourful colour variant appeared at the beginning of the 1980s.

Coloration: Forehead and back appear jade-green. The lower half of the body is light yellow, becoming white on the belly. Head and breast are a red-brown colour, overlain with a light violet tint. Mouth, chin and throat are overlain with light blue. The caudal, pectorals, and ventrals are dark. The grey-blue anal exhibits numerous red dots, and is light violet-blue at its base. There are no egg-spots. The anterior part of the dorsal is red-violet, becoming red posteriorly. The fin is covered in numerous blue streaks and lines, and a very narrow black marginal band completes a very vibrantly coloured dorsal. Females exhibit fewer blue markings and the dorsal may even be completely red. Especially in the northern part of the range, the head may be a striking blue-violet colour below an imaginary line from the eye to the snout, and including the gill region. There are also aberrantly coloured individuals of this local form.

Remarks: My first encounter with this colourful *Tropheus* dates back to a day spent at the "Afrika Aquarium" in Oer-Erkenschwiek near Dortmund. As I later learnt, this was the first consignment to reach Germany. My interest at that time was in fact in other species, but I was sufficiently fascinated by the striking colours of these *Tropheus* to take a closer look. At that time the dorsal of these *Tropheus* appeared to me to be completely red. At that time I was travelling to Dortmund every three weeks in order to see the numerous new imports that were currently arriving from lakes Malawi and Tanganyika. The 600 kilometre journey necessitated immediate decisions as to whether to buy. It was an interesting and wonderful time. In 1993 I bought some tank-bred *Tropheus* sp. "Kasanga". After a few months I was amazed at the coloration gradually developing in these youngsters. The body was covered by a red reticulation. The yellow colour of the lower body changed to yellow-orange. The dorsal became mainly red. I subsequently learned from various sources that these very attractive fishes were very probably the result of a cross between a *Tropheus* "Kachese" and a Red Rainbow.

These sub-adult *Tropheus* were purchased as *Tropheus* sp. "Kasanga" tank-breds. However, as they grew I had my doubts. In all probability they were the result of a cross with a Kachese *Tropheus*. **photo**: P. Schupke

Tropheus sp. Muzi; Population 8.14

Trade name: Tropheus "Muzi", Red Rainbow Blue Blaze, Blue Kasanga Moorii.

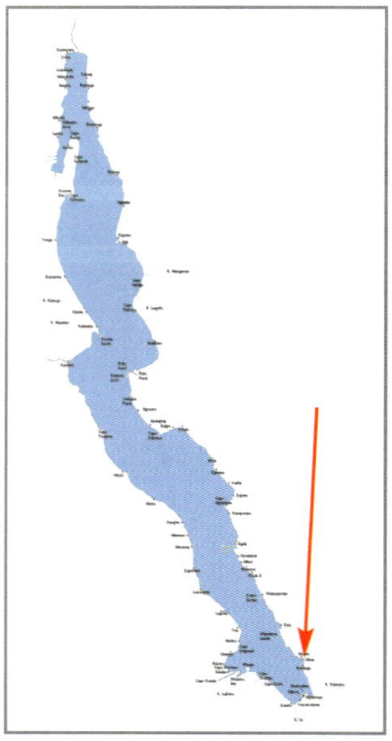

Distribution: DAccording to Ad Konings, the distribution of this local form is restricted to a rocky outcrop at the village of Muzi.

Discovered: This local form was discovered in 1994 by Laif DeMason and Ad Konings during a population inventory in southern Tanzania.

First importation: The first importations for the ornamental fish industry took place in the same year.

Coloration: The base colour consists of a dark green with a dark red reticulation which extends from the head all over the body. There is no lighter patch on the side. Apart from the snout and forehead, the head, breast, and throat are an aubergine colour with a few tiny light blue spots. Depending on mood, the forehead area may assume a turquoise tint. The area between the eyes, the snout, and the posterior corner of the mouth is overlain with violet-bluish. The light blue-turquoise chin stands out to some extent. The dorsal is rather reminiscent of that of *Tropheus* sp. "Kasanga", except that the red is darker.

The blue inclusions are the same. Caudal and pectorals are dark. The anal is light blue at its base, and has a few dark red dots on a blue-grey background. The soft part of the ventrals is likewise dark red. A narrow black marginal band is also present. Aroused individuals may become much darker

Remarks: In the mid 1990s, *Tropheus* sp. "Muzi" was sold as the Blue Blaze variant of the Red Rainbow. The village of Muzi is situated in a bay and is separated from Kasanga by 4 kilometres of sandy beach. A river enters the lake in this area, bringing sediment to the shore in its vicinity. Such areas of shore are avoided by *Tropheus*, and this is evidently the reason for the different coloration of the two populations. It would appear that there is no such clear-cut separation from neighbouring populations to the north, and the coloration changes only slightly there.

Tropheus sp. Molwe; Population 8.15

Trade name: Tropheus "Cherry Rainbow".

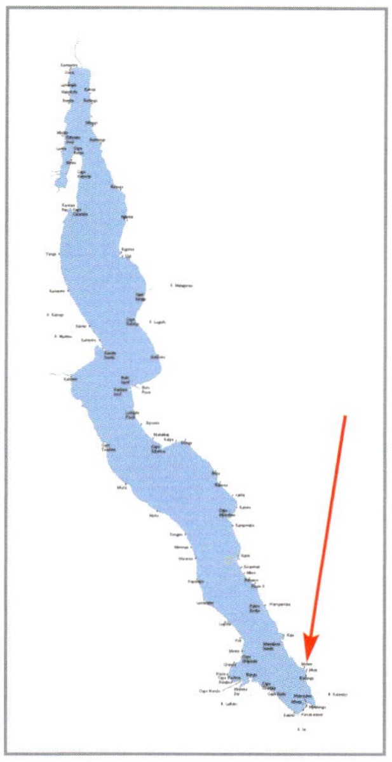

Distribution: The distribution of this local form lies a few kilometres to the north of the preceding form at the village of Molwe.

Discovered: This local form was discovered by Ad Konings and Laif DeMason in 1994.

First importation: As far as I know this form has not been imported into Germany for the ornamental fish industry, but it has been exported for the US market.

Coloration: The forehead and dorsum are a dark green colour, which may, depending on mood, become lighter on the forehead. The rest of the body, and especially the head, is plain dark wine-red. The wine-red varies somewhat according to locality. As in the *Tropheus* from Kasanga, the dorsal is a rather darker red with a few blue markings. Mouth and throat are bluish-green. The caudal, pectorals, and ventrals are dark, but exhibit a dark wine-red tint. There are dark red spots on the membranes of the anal, which is light blue at its base. The vertical barring is often still visible on the dark green area on the back. The iris has orange-red tints in its upper part.

Remarks: The village of Molwe lies in a small bay about halfway between Cape Finga and Kasanga. This is the home of the rather interestingly coloured *Tropheus* sp. "Molwe", which has a very high red component. The American importer Laif DeMason has kindly provided me with a photo which clearly shows the intensity of the colours. If we follow the coastline further to the north then it is noticeable that the red colour is replaced by brown. At Samazi and Katili the *Tropheus* are brownish-green in appearance and have not yet been imported because of their dull coloration. Konings and DeMason have dived along a large part of the coast, but were unable to investigate every single section of shore, so we may hope that one or more additional colour populations may be discovered in the years to come.

Tropheus sp. "Samazi" (not given a population number here) is caught a few kilometres further north of Molwe. Its appearance differs only slightly from *Tropheus* sp. "Molwe".
photo: P. Schupke

Tropheus sp. Mamalesa (Malasa) Population 8.16

Trade name: Tropheus "Malasa".

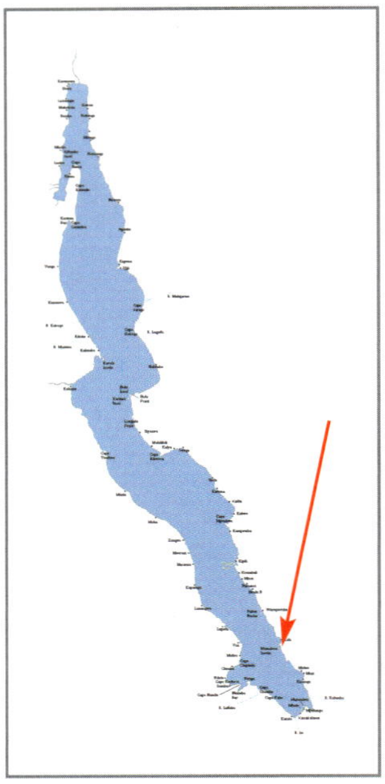

Distribution: The known distribution of this local form is restricted to the Malasa Islands and the nearby coast. Its exact limits are not known.

Discovered: This colour form was discovered by W. EYSEL and myself in the spring of 1983.

First importation: As far as I know, this *Tropheus* was first imported in the mid 1990s.

Coloration: The brown-grey base colour becomes slightly lighter posteriorly and is interrupted by several yellowish vertical bars on the lower half of the body. The shading of the lower stripes is a colour signal that continues north along the coast. Depending on mood, the upper parts of the stripes may also be visible, but they are not as wide as lower down. The head has only a faintly expressed area of colour on the snout and forehead. The dorsal is yellowish-grey, the start of a changing colour trend as we head further north. Caudal and pectorals are very dark, almost black. The anal is variably grey-blue. The eye ring has a yellowish margin, while the rest of the iris has a rather bluish gleam.

Remarks: When we were staying at the Ndole Bay Lodge in Zambia in 1983, we decided to make a trip to the then aquaristically unknown Tanzanian shore opposite. We obtained visas in advance, while still at home. A non-stop trip across the lake is not without its dangers. Strong gales can blow up without warning, sometimes costing the lives of fishermen. Conditions were unfavourable until the last few days of our stay. Armed with Coca-Cola, cold chicken, flare-pistols, spare motor, 40 litres of fuel for the boat, two of the best guards, and copious good wishes, we embarked on our crossing while it was still dark. We would have to start back by 13.00 at the latest, as at this time unfavourable winds could be expected. We thus didn´t have much time to spare. First of all we landed on one of the three Malasa Islands and investigated the fish community there. We then headed for the mainland coast, and an encounter with the customs officials there. This cost us time and patience. We had until midday. The coast there was very precipitous, continuing underwater. We encountered numerous familiar species, nut the local colour forms were new to us. The *Tropheus* were brought up to the boat, studied, and returned to the lake. At that time neither of us was (yet) a *Tropheus* fan. The colours were not very striking and *Tropheus* are not good news for other fishes in transportation containers. We were happy to leave them behind, but we did find some splendid yellow *Ophthalmotilapia ventralis* and an attractive *Cyprichromis* colour variant.

The three Mamalesa Islands off the coast of Tanzania. They are also known as Malasa Islands. **photo**: P. Schupke

Tropheus sp. Kala; Population 8.17

Trade name: Tropheus "Kala Yellow".

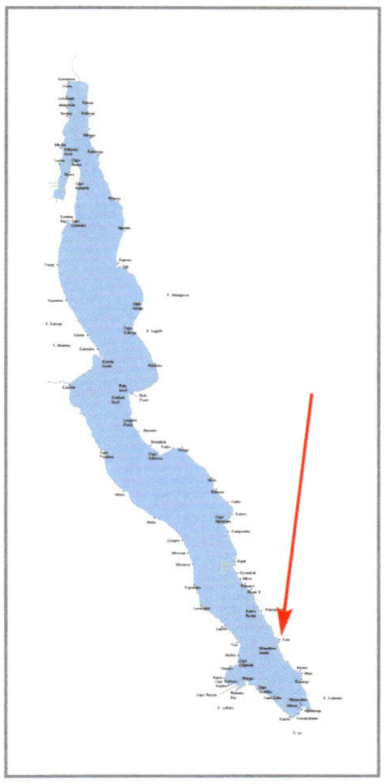

Distribution: The distribution of this local form is said to be in the vicinity of Kala.

Discovered: Ad Konings and Laif DeMason discovered this local form at the Island of Kala in 1994.

First importation: This *Tropheus* was first imported for the German ornamental fish industry at the end of 1994. Further importations have not been made.

Coloration: The base colour is very dependent on the mood of the fish. The dorsal is dark grey-brown, extending over the head to the snout. The violet-brown of the head covers the rest of the body in a reticulated pattern. A yellow patch begins behind the gills on the lower body, extending two scale rows vertically and 7-8 scales horizontally. The snout region, as far as the upper lip, is light blue and the chin light turquoise-green. The remains of yellow vertical bars are often to be seen. Caudal and pectorals are almost black. The dark grey-brown anal may have a few red dots on its spinous part. The strikingly blue-lilac coloured dorsal is bordered with bright yellow at its base; the yellow extends along the rays, creating a striking colour contrast. There is a very narrow black marginal band. In females the dorsal is usually a plain bright yellow, sometimes with blue markings. There are a few tiny light blue dots on the head. In neutral mood the body of this local form is almost plain straw-yellow, and the head too is much lighter.

Remarks: In several places on the Tanzanian coast there are *Tropheus* populations known in which the sexes differ in the colour of the dorsal. These is the case too in *Tropheus* sp. "Kala". The dorsal in females of this *Tropheus* exhibit only shades of yellow. Out of approximately 50 individuals in the importer´s tank, about half had yellow dorsals while in the other half the fin was more or less blue-lilac. The ratio of the sexes must therefore have been in about 50/50. There are also individuals in which there were mere hints of blue in the dorsal. These were smaller males are were dominated by others. The coast north of Kala has not yet been fully investigated, as there is a considerable amount of sedimentation so the habitat is not ideal for *Tropheus*.

Tropheus sp. Wampembe; Population 8.18

Trade name: Tropheus "Wampembe".

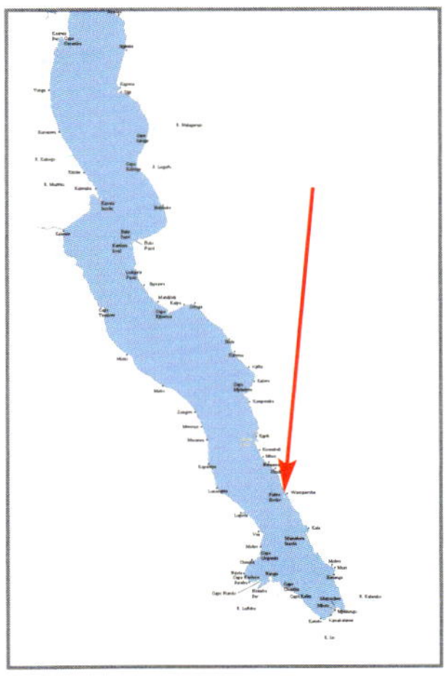

Distribution: The distribution of this colour form lies in the vicinity of the village of Wampembe, at a reef about 3 kilometres offshore.

Discovered: Ad Konings and Laif DeMason discovered this colour form while diving in 1994.

First importation: As far as I know, *Tropheus* sp. "Wampembe" was first imported at the end of 1994. There have been no further imports due to minimal demand.

Coloration: This largely dark-brown fish exhibits very narrow, incomplete whitish vertical bars, becoming broader on the belly. The anterior belly region is whitish-beige. The lower half of the head is a lighter chocolate brown. The dorsal and anal exhibit bluish to greenish tints, and the anal is light grey-blue at its base. The outer ring of the iris is somewhat light blue, while the upper part is brownish. The stripes may disappear, depending on mood. The coloration of older specimens may become much lighter.

Remarks: This local form, which Ad Konings discovered and photographed at Fulwe Rocks, was imported to Germany only once, during the 1994-95 collecting season. The entire consignment was shipped onwards to the USA, where they were supposedly collected by the importer only after a long delay which not all of the fishes survived. A few fishes from this consignment found their way into the hands of scientists, who used DNA study to investigate their phylogenetic position. They established the existence of a close relationship to T. sp. "Chaitika". We will probably have to wait a long time for a further importation as aquarists showed little interest. Konings reports that the shore on the coast at Wampembe and its vicinity consists only of sand with no rocky outcrops. No *Tropheus* are found there. Rocks are not found again until many kilometres further north, at Hinde B. The Wampembe *Tropheus* forms a separate section of lineage 8, which originated from several sub-lineages. This form is also linked to lineage 3, which originated from an ancient main lineage.

Tropheus sp. Hinde B; Population 8.19

Trade name: No trade name is known for this Tropheus colour variant.

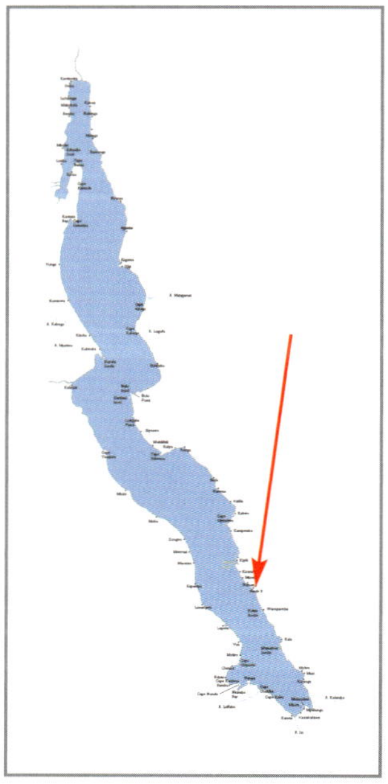

Distribution: The distribution of this colour form is given as the vicinity of the village of Hinde B. According to Ad Konings is the coastline there includes large rocks.

Discovered: *Tropheus* sp. "Hinde B" was discovered in 1994 by Ad Konings and Laif DeMason during an exploratory trip to southern Tanzania.

First importation: As far as I know this *Tropheus* has not yet been imported for the German ornamental fish industry.

Coloration: The base colour is dark brown with irregular yellowish-white vertical bars, which appear rather broader on the lower half of the body. This reinforces the effect of the signal. This brown and white stripe pattern extends a little way into the brown dorsal. Immediately beneath the eye there is a broad vertical stripe which rapidly narrows lower down. The other fins are more or less the colour of the body.

Remarks: This attractive *Tropheus* retains its stripe pattern when adult. It already exhibits a close similarity to lineage 9, which starts just before Kipili, although the local forms of lineage 9 lose their striking barring in adulthood, which is not the case in the *Tropheus* from the Hinde B region. Unfortunately mo aquarium experiences with this colour form are available, so that it is impossible to provide a more accurate description of its various moods and associated colour changes. The sandy coast alternates repeatedly with large aggregations of rocks. Many parts of the coast in this part of Tanzania are bordered by kilometres-long sandy beaches and provide no habitat for Aufwuchs-feeders. The rocky shoreline does not resume until further north. It would, however, be nice to see this *Tropheus* colour form imported in order to learn about its requirements and perhaps find enthusiasts willing to work with these fishes in the long term.

The *Tropheus* from Hinde B are said to retain their striped pattern throughout their lives. No aquarium. observations of this interesting form are available.

photo: A. Konings

Tropheus sp. Namansi I; Population 8.20

Trade name: Tropheus "Namansi", Red Namansi.

Distribution: To date this local form has been reported only from the village of Namansi in Tanzania. The precise limits of its distribution are unknown.

Discovered: This local form was discovered Ad KONINGS and Laif DEMASON in 1994 during their investigation of the coast in southern Tanzania.

First importation: This colour form was imported for the German ornamental fish industry at the end of 1994.

Coloration: The base colour consists of a greenish grey-brown, overlain with a brown-red reticulation, particularly on the anterior part of the body. The brown-red head and throat are overlain to some extent with bluish-grey on the snout, mouth, and chin. The lower half of the body is somewhat lighter. The whitish-yellow vertical bars are still visible in adults, and are visibly wider and more prominent in their lower half. The upper part of the eye is yellowish. The yellowish-grey dorsal has a very narrow black marginal band. Caudal, anal, pectorals, and ventrals are very dark in colour, and have no special colour features except for the anal, which is bordered light blue at its base. Depending on mood, these fishes can become darker or a lot lighter.

Remarks: The *Tropheus* from Namansi area exhibit a degree of similarity to the Rainbows on the other side of the lake at Mwerazi in the former Zaire. I have been unable to gain any personal experience with *Tropheus* sp. "Namansi", as the few fishes that I was able to observe at the importer´s had already been sold. At first glance I thought this local form was *Tropheus* "Murago", except that these fishes lacked the dots on the head. Female Murago also have a yellow dorsal. As with *Tropheus polli*, the fishes were larger on average than other, familiar *Tropheus* imports. The above-average size of this colour form may be the result of better environmental conditions or simply a function of the collecting technique used. According to KONINGS, the distribution of lineage 9 begins at Namansi. However, these *Tropheus* live at rather greater depths than the *Tropheus* of lineage 8.

Tropheus sp. Mtosi I; Population 8.21

Trade name: I know of no trade name for this local form.

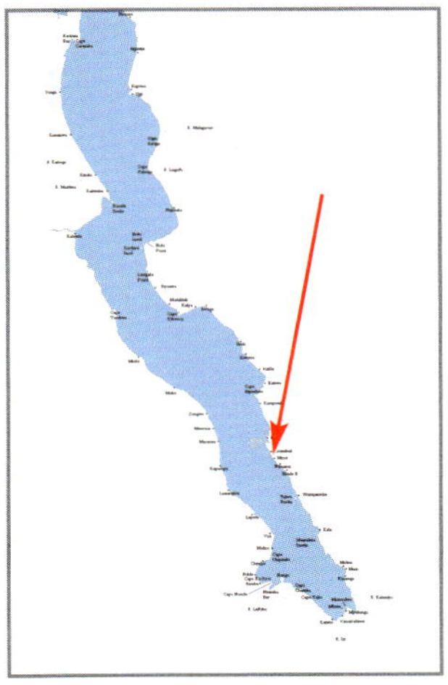

Distribution : This local form is recorded as being found at Mtosi (M´Tossi). The precise limits of its distribution are unknown.

Discovered: *Tropheus* sp. "Mtosi" was discovered by Ad KONINGS and Laif DEMASON in 1994 during their investigation of the area.

First importation: These *Tropheus* have not yet been imported in significant numbers for the German ornamental fish industry.

Coloration: The base colour of the body is divided into two zones: an upper brownish-green area and a lower yellowish-white one. The brownish head has no dots. The slightly yellowish-white vertical barring is bolder on the belly region. The stripes are retained in adult specimens, and continue a short distance into the brownish-yellowish dorsal, which has a narrow black marginal band. The caudal, like the pectorals and ventrals, is very dark, but has a narrow lighter terminal zone. The anal exhibits brownish-red tints. The eye is silvery-bluish with brownish-red in its upper part.

Remarks: *Tropheus* of lineage 9 are also found in the vicinity of Mtosi. The Rainbow *Tropheus* are said to be sympatric with lineage 9 there, but at different depths. Localities are known in other parts of the lake where *Tropheus* of different lineages encounter one another, but insufficient research has been conducted into whether and to what extent these different lineages overlap. The *Tropheus* of lineage 9 live at appreciably greater depths, while the members of lineage 8 inhabit the upper part of the rocky littoral. *Tropheus* sp. "Mtosi" prefers sediment-free substrates, a preference shared with their relatives of the same lineage in Zambia. It is also interesting to note that a few *Tropheus* of this lineage retain their barring as adults, while those of lineage 9 almost of completely lose their striking stripes when adult.

Tropheus sp. Kisambali; Population 8.22

Trade name: I know of no trade name for this local form.

Distribution: This local form lives a few kilometres north of the Mtosi *Tropheus*, and is found at Kisambali.

Discovered: Ad Konings and Laif DeMason discovered this colour form during their explorations in 1994.

First importation: As far as I know, no specimens have been imported for the German ornamental fish industry. It is possible that small numbers may have been brought back privately.

Coloration: The base colour of the body is brown-olive with light vertical bars that broaden on the lower half of the body, creating the impression of a fragmented lateral spot. Depending on mood, the body may become lighter and is then more of a light greenish-brown contrasting with the yellowish grey-brown dorsal. The bluish-grey anal has a few orange-red dots on its spinous part. The dark caudal has a lighter terminal zone. The yellowish caudal tips are rather distinctive. The lower part of the eye ring has a rather bluish-grey sheen, while the upper part is brownish-red.

Remarks: This local form is the most northerly representative of lineage 8 in Tanzania, and (as far as is known) the distribution of the lineage ends here. This form too is sympatric with *Tropheus* of lineage 9 in this section of coast. The *Tropheus* of the preceding local form, at Mtosi, and this local form differ little, at least in neutral mood, according to the underwater photos of Ad Konings. An interesting point is the similarity in external appearance between the two lineages 8 and 9, which are found sympatric for several kilometres, albeit at different depths. It would appear that barring - whether prominent on the lower half of the flank or visible to a greater or lesser extent all over the body - is the most successful pattern of markings in this area of coast. Scientists term this phenomenon convergence.

Tropheus of lineage 9

The first *Tropheus* of this lineage arrived in the German ornamental fish industry back in the mid 1980s. At that time they were sold as "Wasp Moorii" and available only in very small numbers. Thereafter there was a gap of several years. Not until the end of the 1980s were several, and at the mid of the 1990s all known, populations offered by the relevant importers. We have Ad Konings and Laif DeMason to thank for our knowledge of these colour forms, following their exploration of the southern coast of Tanzania in 1994. At that time they discovered six well-defined populations. Although they explored a large section of coast, they had to limit their researches to those stretches of shore that appeared suitable for *Tropheus*. Hence the possibility remains that a few colour variants may have been overlooked along the long sandy coast.

The distribution of lineage 9 begins about 30 kilometres south of Kipili, at Namansi in Tanzania, and extends - as far as we know at present - to Karema, a few kilometres south of Ikola. The distance between Karema and Namansi is about 100 kilometres, and includes a number of very long sandy shores and river mouths. In the southern part of its range lineage 9 lives sympatric with *Tropheus* of lineage 8, the "Rainbows", for several kilometres. No comparable data are available for the northern area at Karema. To the north of Karema the river Ifume enters the lake, the inflowing water creating a large area that is avoided by Aufwuchs-feeders and representing an insuperable barrier for often long periods of time. A few kilometres to the north lies the village of Ikola, 20 kilometres to the north of which we find the southernmost limit of the distribution of *Tropheus* sp. "Ikola", better known as the Kaiser Moorii (lineage 11). At present it is not known whether there is any contact between the two lineages. The *Tropheus* of lineage 9 prefer deeper water in which a certain amount of sediment has been deposited. In the overlap zone to the south the Rainbow *Tropheus* occupy the upper rocky littoral where there is no sediment. Interestingly, the two lineages are very similar in their coloration in that area, probably an adaptation to the prevailing conditions rather than indicating any close relationship between the two lineages.

The *Tropheus* of this lineage differ very clearly in several ways from all the other lineages in Lake Tanganyika. All populations have a light blue iris. Non-dominant and sexually inactive adults, and juveniles, have a distinct pattern of stripes on the body. Dominant adults, usually males, lose their stripes (almost) completely, although in situations of severe stress the bars again become clearly visible. In *Tropheus* "Fiery Fry" from the Namansi region, recently imported for the first time, both sexes are the same colour. In *Tropheus* of lineage 9 the tail is almost straight-edged, and all the fins are the same colour. Juveniles have a very striking, contrasting colour pattern. They are brown-black with vivid red-orange, yellow, or whitish yellow vertical stripes of uniform width. In the lake this is an exceptional camouflage pattern, but in the aquarium this combination of colours makes these fishes highly desirable. Unfortunately the vivid colours are gradually lost with increasing age. Ad Konings reports that the individual colour forms are sometimes linked by intermediate forms. There is an unmistakeable similarity to *Tropheus polli* from lineage 4, which penetrates into the northern distribution of *Tropheus* sp. "Ikola". Leaving aside the completely different tail shape, juveniles exhibit an almost identical pattern of colour and markings. Even the colour of the iris is the same. It is often asserted that *Tropheus* of lineage 9 are more aggressive than *Tropheus* of other lineages, but I have never been able to confirm this observation from the colour forms I have kept myself. Possible reasons for the problem include an incompatible group or single individuals that need to be removed in order to create harmony within a group. The fishes of this lineage have 6 anal spines.

This lineage contains the populations numbered 9.0 to 9.8

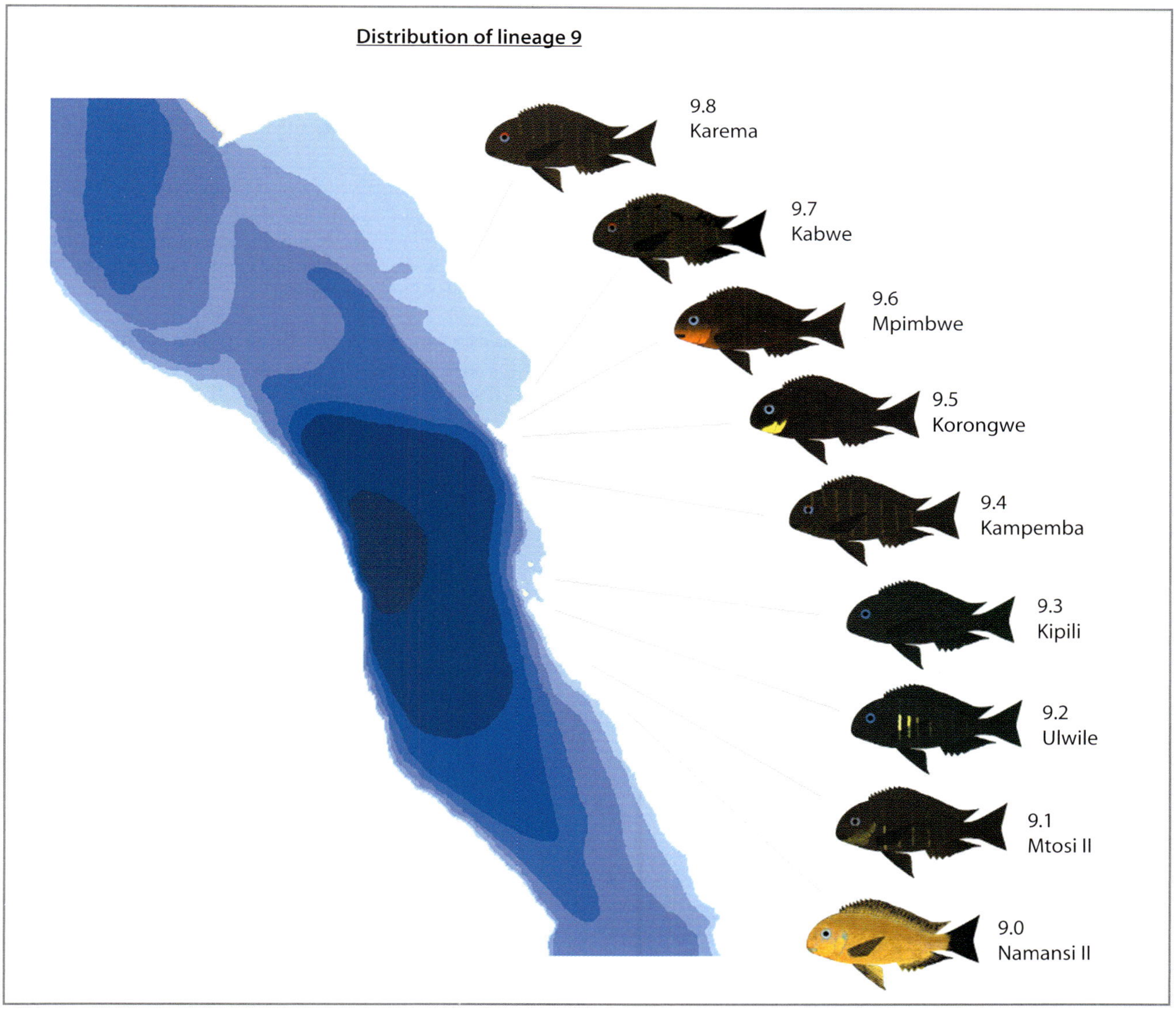

Tropheus sp. Namansi II; Population 9.0

Trade name: Tropheus "Fiery Fry"

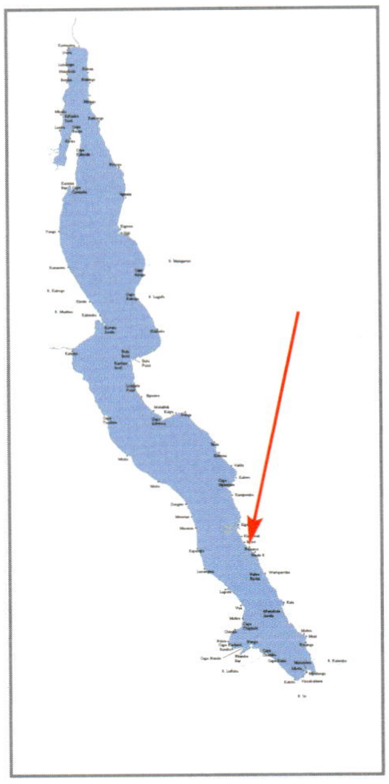

Distribution: The distribution of this population lies in the vicinity of the village of Namansi in Tanzania.

Discovered: This *Tropheus* was probably discovered in 1994 by Ad Konings and Laif DeMason during their exploration of the southern Tanzanian coast.

First importation: According to my information, this population was first offered for sale in Germany in 2000.

Juvenile

Mutant

Coloration: In unstressed individuals the base colour consists of an orange-brown, which may be rather darker and greenish in the dorsal, and becomes vivid orange-yellow on the underside. The area below the eye is likewise bright orange-yellow. The iris is white-blue. There are bright pink-coloured areas around the eye, in the gill region, and at the base of the pectorals. Sometimes light vertical bars are faintly visible. Both sexes are the same colour, at least in neutral mood. Young fishes exhibit a very striking juvenile which diminishes with the onset of sexual maturity and eventually disappears completely: they have a dark brown or, depending on viewpoint, bright red background, with vertical bars of the other colour. In 2001 a mutant form of this population was discovered, but unfortunately I have seen this fish only in a photo on the Internet. This individual was yellow-orange on the body while a small area on the back and the part of the dorsal above it were dark in colour, likewise the caudal and a narrow marginal band in the dorsal.

Remarks: According to our current knowledge, this population is the southernmost representative of lineage 9. It escaped discovery for a long time die to the depth of its habitat. It lives on the same stretch of coast as *Tropheus* sp. "Namansi", population number 8.20, with the *Tropheus* of the "Rainbow lineage" preferring the upper part of the rocky littoral. This phenomenon is known from other parts of the lake where in some cases up to four *Tropheus* species share the biotope.

Tropheus sp. Mtosi I; Population 9.1

Trade name: Tropheus "Mtosi"

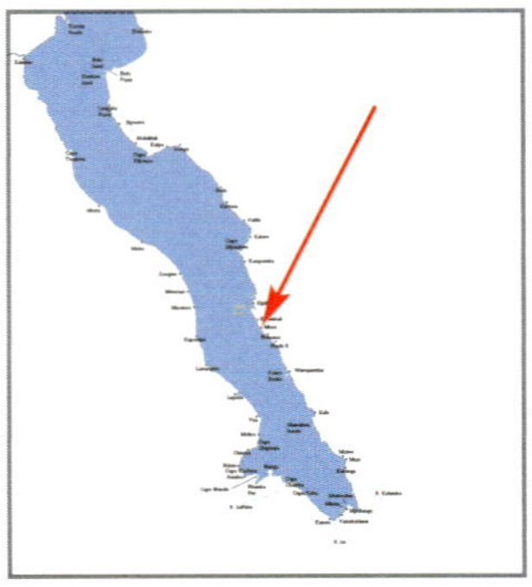

Distribution: The distribution of this local form is said to be the region around Mtosi (or Mtossi, as it was formerly known).

Discovered: This *Tropheus* was discovered by Ad Konings and Laif DeMason during their exploration of the southern Tanzanian coast in 1994.

♂

First importation: This *Tropheus* has been imported only in small numbers for the German ornamental fish industry.

Coloration: The base colour consists of a brown-green mixture, which, depending on mood, may darken or lighten. Vertical bars are only indistinctly visible. The first stripes behind the gills are rather more prominent and pale yellowish in colour. Further back the intensity of the stripes decreases. The region of the posterior and lower gill-cover likewise exhibits shades of a rather lighter yellowish-green. The fins are the same brown-green as the body. The caudal has an almost straight posterior edge. The iris is light blue. There is a rather lighter area between the eye and the snout. Juveniles have an attractive orange-yellow and dark brown vertical barring. They are indistinguishable from juvenile *Tropheus* from Kipili. Unfortunately they lose their bright colours very rapidly as they gradually change to adult coloration.

Remarks: It is not known whether all representatives of lineage 9 are found n deeper water or only where they encounter *Tropheus* of lineage 8. A possible reason for this preference may be that the Rainbow *Tropheus* are better able to compete for habitat, or alternatively it may be that the *Tropheus* of lineage 9 are a more ancient lineage and their evolutionary potential is less than that of the putatively younger *Tropheus* of the "Rainbow lineage" in this region. Once again there is a serious shortage of information regarding the origins of the numerous lineages. Only one thing is certain - Lake Tanganyika has experienced several epochs, and the divisions within the genus *Tropheus* reflect these periods. If we could discover the approximate age of the individual lineages, then it would be rather easier to hypothesise regarding the competition between them.

Tropheus sp. Ulwile; Population 9.2

Trade name: Tropheus "Ulwile"

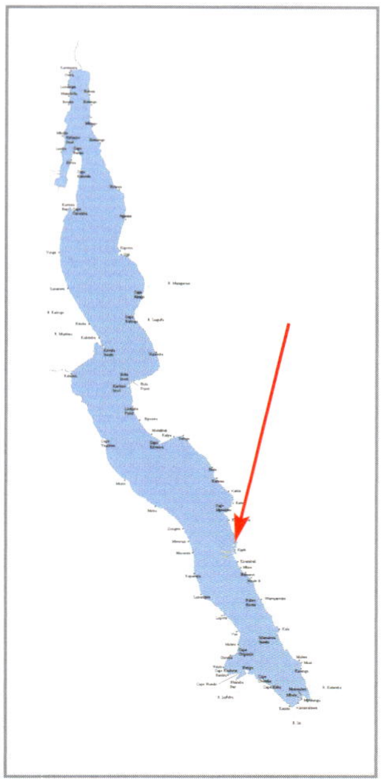

Distribution: The distribution of this local colour form is at the Island of Ulwile, which lies a few kilometres south of Kipili.

Discovered: In 1994 Ad Konings and Laif DeMason investigated this area, with the goal of understanding the local cichlid populations and introducing them to an interested public.

First importation: This local form was, as far as I know, first imported for the ornamental fish industry following its discovery during the 1990s.

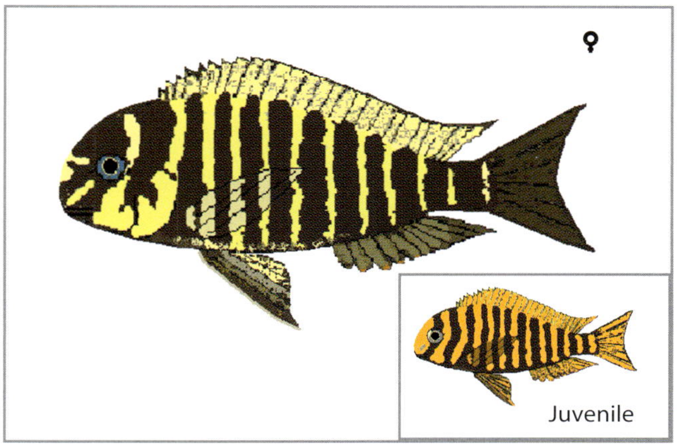

Coloration: The base colour consists of an olive-black, which, depending on age and motivation, may exhibit yellow vertical bars of variable intensity. Sexually active males lose their stripes apart from residues of the 3-4 bars on the central flank behind the gills; their bright yellow colour contrasts boldly with the otherwise dark body. Females have yellow stripes and patches all over the body and head respectively. Juveniles have striking orange bars, but lose this lovely coloration in the course of time, changing to a yellow shade in males and grey-yellow in females. The light blue iris of adults is not so prominent in juveniles - the light blue develops and is at its most prominent only in adults. In females the dorsal is a yellowish colour, while in males it looks rather bluish and becomes dark when the fish is aroused. The head is pink as far as the posterior operculum. Throat and breast have rather yellowish areas.

Remarks: These *Tropheus* are a dramatic sight in the aquarium. The strongly contrasting yellow too grey-yellow vertical bars, the light blue eyes, and the pink head make for a *Tropheus* that compares favourably with other forms where colour is concerned. Juveniles are interestingly coloured, with bright orange bars on a dark brown background but unfortunately, like *Tropheus duboisi*, they change to an adult coloration which, as in many other *Tropheus*, loses its intensity of colour. The *Tropheus* of population 9.2 were not maintained in the aquarium for very long, as apparently there were problems with their maintenance. Nowadays they are practically unavailable. A tank-bred population was never established and most of the imported fishes are no longer alive.

Tropheus sp. Kipili; Population 9.3

Trade name: Tropheus "Kipili", Wasp Moorii

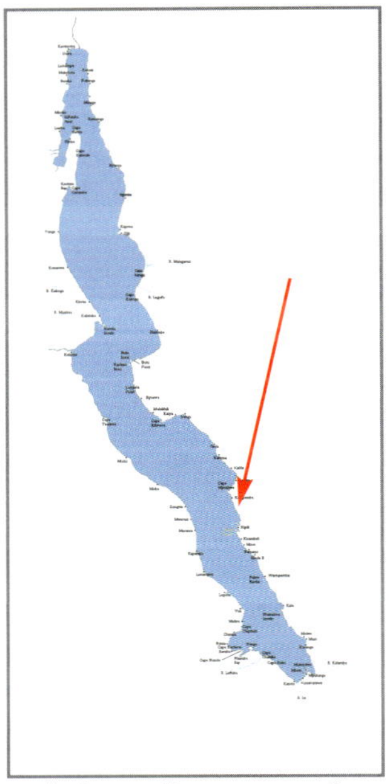

Distribution: The distribution of this colour form is in the general area of Kipili.

Discovered: I have no information on who discovered this local form in the mid 1980s.

First importation: In the early 1980s a colour form with the name "Wasp Moorii" was imported in only small numbers.

Coloration: The base colour of the various age-related stages is very different. Juveniles have a bright orange body and fins, with dark brown vertical bars. With increasing age the orange colour is lost, changing gradually to yellow. Females and, depending on mood, males retain this pattern, including the colours. Dominant adult males are olive-black all over. A few individuals of this local form have yellow stripes behind the gills. Apart from the anterior part of the dorsal, which may be somewhat lighter in colour, all the fins are the same dark colour as the body. The dorsal of inactive males exhibits a slight light-blue tint, changing to yellow posteriorly. The fin has a narrow, barely visible, marginal band. The iris is light blue.

Juvenile

Remarks: DThe ability of sexually active males of lineage 9 to use their stripes as a signal is most pronounced in the Ulwile

This only slightly aroused male still shows residues of his stripe pattern. The yellow stripes on the central flank increase in intensity of colour.

The same male in neutral mood. His appearance is quite different. The rather bluish colour in the dorsal is now more apparent.

photos: P. Schupke

population. This characteristic diminishes rapidly in the south as well as to the north, and eventually disappears completely. The situation is rather different in females, which exhibit yellow coloration throughout much of the range of lineage 9. Except in the area to the north of Cape Mpimbwe, juveniles exhibit a more or less comparable pattern of markings throughout the range. North of Cape Mpimbwe juveniles lose the yellow more and more, and their stripes are pale yellow to whitish.

Tropheus sp. Kampemba; Population 9.4

Trade name: Tropheus "Kampemba"

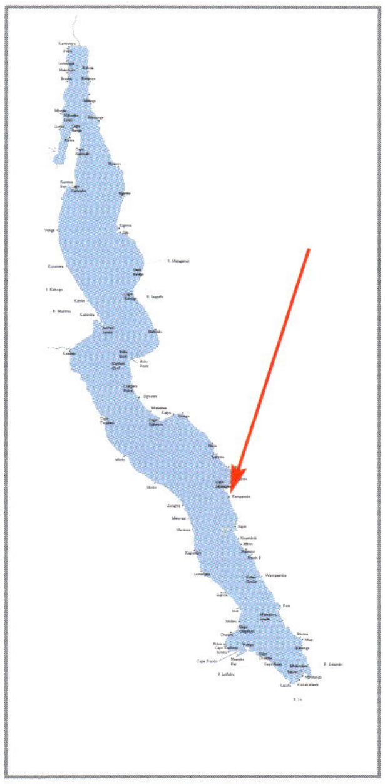

Distribution: The distribution lies in the general area of Cape Kampemba; its precise limits are unknown.

Discovered: The *Tropheus* of this local form were discovered by Ad Konings and Laif DeMason during their exploration of this area of coastline in 1994.

First importation: Small numbers of this local form were imported for the German ornamental fish industry in 1995.

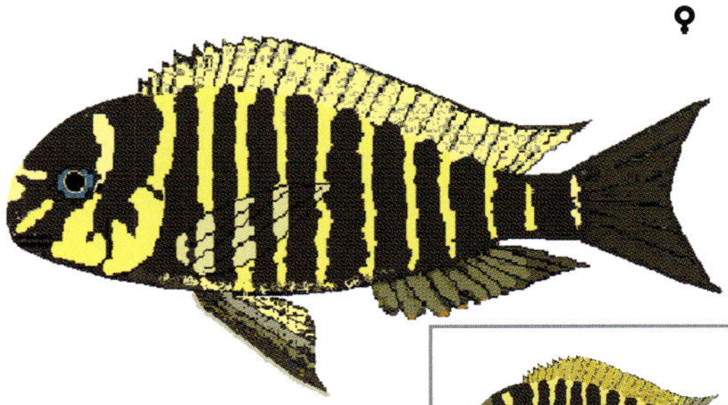

Juvenile

Coloration: As already seen in the neighbouring population, the sexes exhibit different colours depending on mood. Dominant adult males are a uniform dark brown without any markings or colour pattern; only the pectorals have a dark yellow tint between the rays. The cheeks are somewhat lighter with something of a dark pinkish sheen. In neutral mood pale vertical bars are visible, without there being much colour contrast. Females, by contrast, have a striking pattern of dark brown and broad yellow stripes. The lips are blackish. In males the rest of the fins are the same dark colour as the body. To make sense, that sentence should follow "yellow tint between the rays" above. The anal may have orange tips to the spines, with a few small pale egg-spots on its posterior part in males. The iris is light blue.

Remarks: In this local form the base colour of males changes from the olive-black seen at Kipili and the offshore islands to a dark brown colour which continues to the northern limit of the lineage. In females and juveniles a change is apparent just a few kilometres north of Cape Mpimbwe. The rather yellowish coloration of the pectoral fins and the fact that the southern coast of Tanzania has not yet been fully explored lead to the hope that a local form will be discovered that has brighter yellow pectorals than in the *Tropheus* at Kampemba.

In aroused males the vertical bars disappear. The yellow colour between the rays of the pectorals is readily visible. The cheek region is somewhat lighter, faintly reminiscent of the neighbouring population at Cape Mpimbwe.

photo: P. Schupke

Tropheus sp. Korongwe; Population 9.5; *Tropheus* sp. Mpimbwe; Population 9.6

Trade name: Tropheus "Mpimbwe", Tropheus "Dahabu", Tropheus "Msalaba".

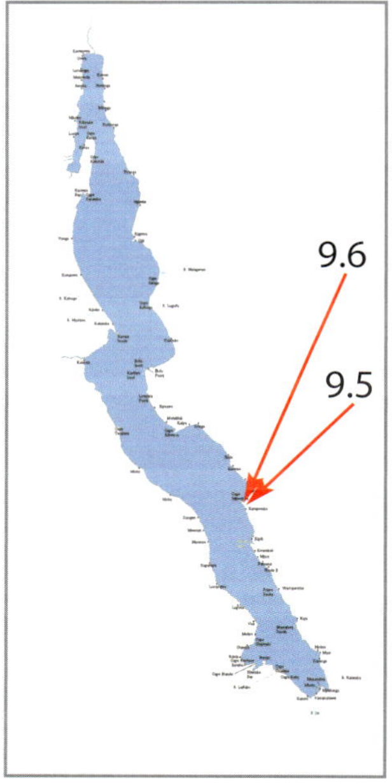

Distribution: The distribution of this local form is only inadequately known. Population 9.6 inhabits an approximately 5 kilometre rocky coast that includes Cape Mpimbwe, known locally as Msalaba, and the bay. Population 9.5 lives to the south. The two forms are probably very closely related and it is likely that intermediates occur.

Discovered: No details are available regarding who discovered this local form, and when.

First importation: As far as I know, these *Tropheus* were imported for the ornamental fish industry at the end of the 1980s.

Coloration: The base colour of dominant adult males is a dark brown-red without vertical bars. Depending on mood, the stripes may reappear. This occurs mainly following loss of territory or transfer to an unfamiliar tank. Males then look like females, except that they have a high orange component in their stripes. Beneath an imaginary line from the corner of the mouth to the lower edge of the eye and the opercular spot, dominant males of population 9.6 exhibit an orange to orange-yellow colour, which changes to yellow in population 9.5 a few kilometres to the south. The extent of this coloured zone varies with the individual. Some specimens have only a narrow orange-red bar on the lower gill-cover region, while in others the coloured area extends almost to the eye. The females of the two populations look the same and have straw-yellow vertical bars that become white towards the back, and the orange facial zone seen in males is straw-yellow to whitish. All the fins are an inconspicuous, indefinable, rather translucent mix of grey and brown. The barring extends a little way into the base of the dorsal. The iris is light blue with brown-red in its upper part.

9.5, Korongwe, ♂

9.6, Mpimbwe, ♂

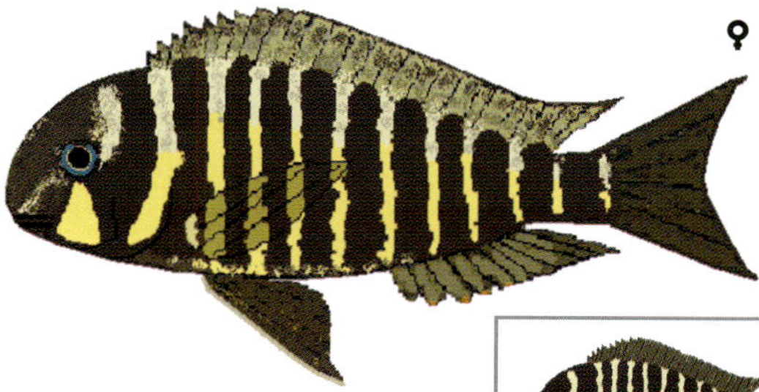
♀

Juvenile

Cape Mpimbwe (Msalaba) is noted for its large blocks of rock above and below water. In the background can be seen the mountain ridge at whose left-hand end lies the village of Karema.
photo: H. J. Herrmann

Remarks: The eventual full-grown size of these local *Tropheus* forms is barely 20 cm. An eventual size of this type is otherwise seen only in *Tropheus* of lineage 4, to which *Tropheus polli* belongs. Other *Tropheus* do not attain this size. Fishes of this size require correspondingly large territories, which can easily exceed the 2 metre diameter mark. This means a 2 metre aquarium is rather on the small side. A large number of individuals will reduce territory size, but how many people can afford a group of 30 or 40? A smaller tank with fewer inhabitants is one solution, but is not suited to the requirements of the fishes.

Tropheus sp. Kabwe; Population 9.7

Trade name: Tropheus "Kashekezi", Tropheus "Golden Kushangaza".

Distribution: Kabwe is situated about 10 kilometres north of Cape Mpimbwe (Msalaba).

Discovered: This population was discovered in 1994 by Ad Konings and Laif DeMason. The mutant form was not discovered until a few years later when this stretch of coast was examined more closely.

First importation: As far as I know, the normal form reached Europe for the first time in 2000. The mutant form was offered for sale in autumn 2001 at Helmut Löfflad´s "Cichlid Shack" in Bühl, Bavaria. Whether this form had already been imported previously is beyond my knowledge.

Coloration: The entire fish is normally dark brown. Depending on mood, it may become rather lighter, or darker when aroused. Sometimes lighter vertical bars can be seen. Sometimes odd black spots may appear on otherwise normally coloured individuals. The coloration of females is as yet unknown. That of juveniles is identical with that of juveniles of population 9.3. at Kipili. The blue iris is barely noticeable and this population shows an element of red in the upper iris. In this

Lighter vertical bars appear under stress or domination by stronger males. The black spots in the posterior dorsal indicate defects in the melanin distribution.
photo: P. Schupke

colour variant mutant forms occur in which patches of the skin lack melanin so that an orange-yellow colour appears. The distribution of such markings varies considerably from individual to individual.

Remarks: According to KONINGS, this species occurs in very shallow water with very large rocks. Visibility is very poor year-round. In autumn 2001 I was able to take a closer look at a number of imported specimens of the mutant form at Helmut LÖFFLAD´s "Cichlid Shack". They did not look particularly attractive, but they had only just arrived. The melanin-free areas were very small on these fishes, and the correspondingly large amount of black rather spoiled the effect. Especially attractive specimens are probably reserved for customers who are better off, but this too often goes hand in hand with lack of experience and these jewels are usually lost to the hobby. It remains to be seen what will happen when these melanin-reduced individuals are bred. The genetic inheritance of these mutants is so far unknown (or, at least, has not been publicised). Another melanin-reduced mutant, the "Golden Kalambo" (a mutated form of population 8.12, no drawing, see photo section of this book), which is likewise high-priced, and likewise has yet to be cultivated in captivity. These fishes are correspondingly rarely available in the trade.

Tropheus sp. Kalila; Population 9.8

Trade name: Tropheus "Kalila Red Eye", Tropheus "Mkombe", Tropheus "Karema".

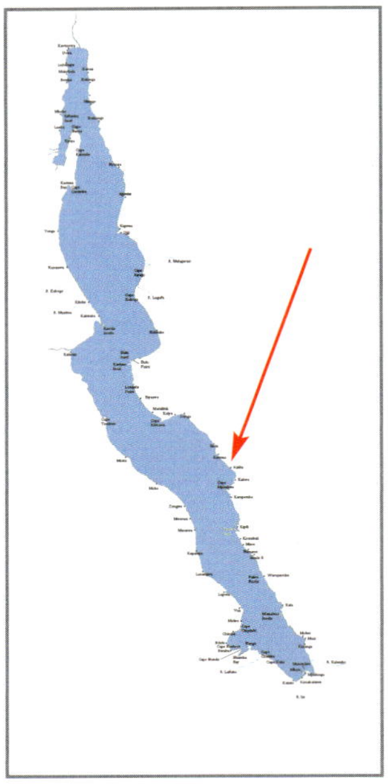

♂

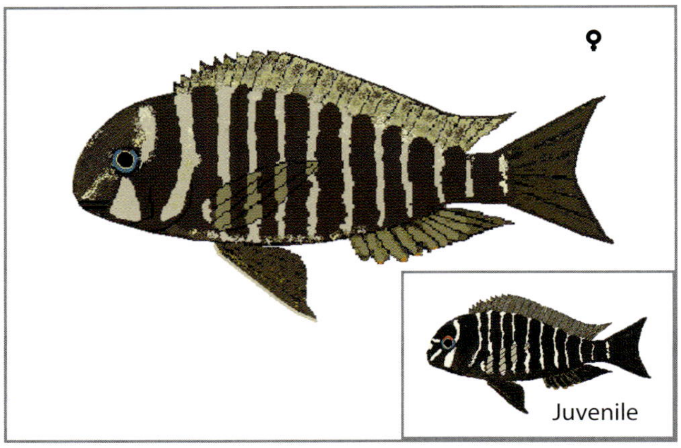

♀

Juvenile

Distribution: The distribution of this local form extends for a few kilometres south of Mkombe, which lies a few kilometres north of Cape Mpimbwe, to the village of Karema. That represents some 35 kilometres of coastline.

Discovered: No details are available regarding the discovery of this local form.

First importation: As far as I know, this form has not been imported for the German ornamental fish industry. It is, however, quite possible that a few individuals were imported in the mid 1990s. These *Tropheus* have been imported into the USA.

Coloration: The base colour of sexually active males is uniform dark brown. They no longer exhibit vertical barring. Juveniles and females have narrow white vertical bars, only a little wider on the central flank, on a dark brown background. Otherwise the markings are identical with those of other populations. Depending on mood, both sexes may become much lighter. The slightly darker dorsal and anal are translucent. Caudal, pectorals, and ventrals are very dark in colour. The light blue iris has a striking red area in its upper part.

Remarks: The *Tropheus* of this local form inhabit a 35 kilometre long stretch of coastline. Individuals from Mkombe, the southernmost location for this population, resemble those from Karema. It is conceivable that there are nevertheless small variations between these two locations. As far as is known at present, the distribution ends at Karema. North of the village is the broad estuary of the River Ifume, and this at present represents a distributional barrier that prevents movement by *Tropheus* in either direction. North of this area of swamp and reed beds lies the southern limit of the distribution of *Tropheus* sp. "Ikola", which belongs to another lineage (11). The area south of the Ifume river is inadequately studied. Perhaps closer study will reveal one or more surprises.

Tropheus of lineage 10: comprises *Tropheus duboisi* Marlier, 1959

Scarcely another *Tropheus* has captured the hearts of aquarists world-wide as much as *Tropheus duboisi*. Its striking and highly contrasting juvenile coloration is no less attractive than that of adults. In addition, its behaviour towards conspecifics is far less aggressive than in other *Tropheus* species. It has been firmly established in the aquarium hobby for many years and there can be hardly anyone who has not fallen under its spell. Often a first encounter with these cichlids leads on to cichlids in general, and many years of involvement with these interesting fishes. *Tropheus* of lineage 10 have a wider distribution that was thought when *T.*

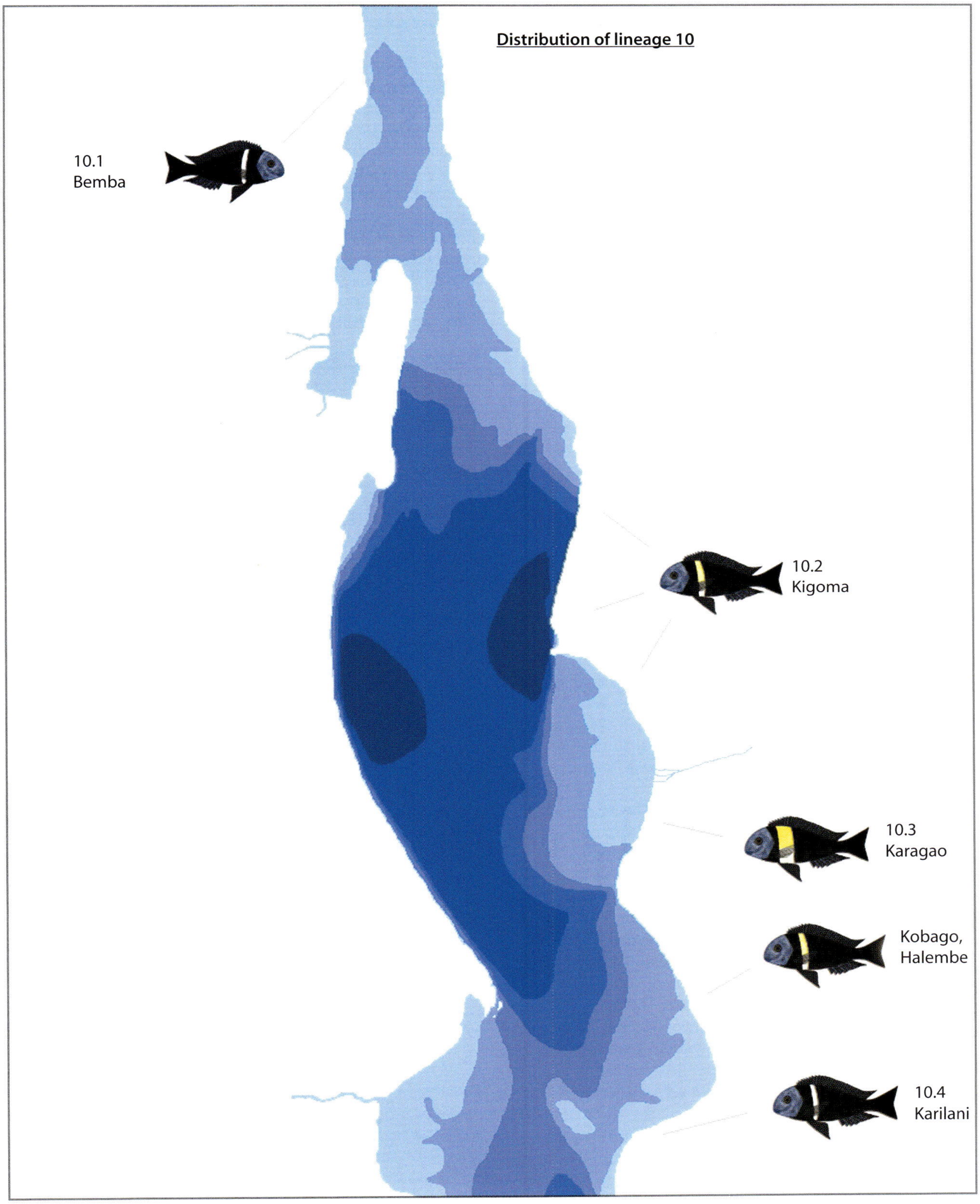

duboisi was first described in 1959. Over the years new populations have repeatedly been discovered. The first known distribution area lies in the north-western corner of the lake at Cape Munene in the Congo, where it lives sympatric with *Tropheus* sp. "Bemba" (Pop. 2.4) along a short section of coast close to the the village formerly known as Bemba. Further populations were subsequently discovered between Nyanza Lac and Bulu Point. The largest continuous population is found on the east coast and extends from Nyanza Lac south to Halembe, which lies south of the mouth of the Malagarasi. Here the lateral stripe in these fishes, which varies in width, exhibits a more or less yellow colour. A new population was discovered a few years ago at Karilani Island, which lies close to the shore north of the lofty Mahale Mountains. It resembles the northerly population at Bemba.

The appearance of *Tropheus* of lineage 10 is the same in all collecting localities apart from the width and colour of the stripe. At all locations they are sympatric with other *Tropheus* species, but do not always come into immediate contact with them, as their preferred habitat - 3 -15 m of depth - is below that of the other *Tropheus* species.

Young *Tropheus* of lineage 10 have a very different appearance to their congeners. They are completely black with numerous gleaming white dots which are arranged in barely perceptible vertical rows. This colour pattern makes them extremely desirable aquarium fishes all over the world. At the age of several months the dots gradually disappear and a white or yellow band begins to become visible beneath the anterior spines of the dorsal. By the time all the spots have disappeared the band is usually complete. There is an element of sexual dimorphism, in that adult males have longer ventral fins, extending back to the anal.

According to Sturmbauer and Meyer, *Tropheus duboisi* is one of the oldest *Tropheus* lineages, supposedly am offshoot of the original ancestor of all *Tropheus*. It has been suggested by other authors that their extreme age, compared that of other, younger, more adaptable *Tropheus* lineages, has led to their being forced into deeper water or completely eradicated in some areas. Another possibility is that over the ages *Tropheus duboisi* has adapted to the food supply at greater depths and found its specific niche there. Between Nyanza Lac and Cape Kabogo the population density is said to be rather high, which does not accord so well with the theory of suppression by younger species.

Brichard hoped to find additional *Tropheus* of lineage 10 populations along the coast between the Ubwari peninsula and Kalemie in what was then Zaire, but was unable to do so. The area south of the Ubwari peninsula as far as the village of Kunanwa is inhabited only by *Tropheus* of lineage 5. This is one of the oldest coasts of the lake and is thus theoretically a suitable location for *Tropheus* of lineage 10.

In the natural habitat of the rocky littoral *Tropheus* of lineage 10 never remain for long at any one spot. They are constantly on the move. They move, feeding, around the larger blocks of stone and repeatedly disappear from view. Adults are generally solitary. They can be seen in pairs when they are mating.

Four populations within the lineage are distinguished below. It is impossible to comment here on the possible existence of additional populations that lack the light band. I know these only from photos more than 30 years old. No locality for such fish is known in the wild, nor have there been any recent sightings reported.

Tropheus duboisi Marlier, 1959; Population 10.1

Trade name: Tropheus duboisi "White Band", White Band Duboisi, Bemba Duboisi

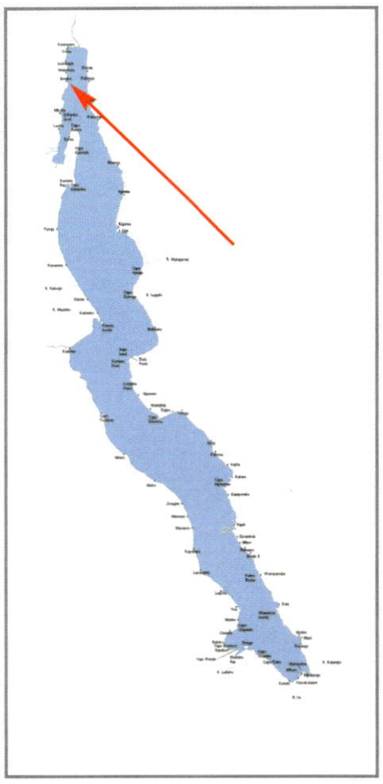

Distribution: The distribution of this local form lies along a barely 1 kilometre stretch of coast at Cape Munene, near to the site of the former village of Bemba.

Discovered: The *Tropheus duboisi* from the former Bemba was probably discovered in 1958 and described a year later by Marlier.

First importation: This local form was first imported for the ornamental fish industry 1958. However, these specimens died out after a short time and the species was not imported again until 1974.

Coloration: The black of the body is interrupted by a snow-white vertical band, 3-5 scales wide, that begins beneath the 3rd to 4th dorsal spines and extends under the belly. This band can be variable in its prominence - there are individuals in which it is virtually absent or represented by just a few white scales where the band would be, if present. All the fins, and particularly the head, are overlain with a blue-whitish colour. The colour of the band may extend a little way into the dorsal. The anal has no egg-spots or the like. The iris is dark. Juveniles are very different in appearance to adults. They too are black all over, but have gleaming white dots all over the body as well as the dorsal and anal. On closer inspection these dots can be seen to be arranged in vertical rows. When the fry leave the mouth for the first time, it is possible to discern yellowish colour between the white dots, and this may be the vestiges of former (yellowish?) vertical bars that the species may have possessed in the past. Over the months the spots gradually disappear and the future band slowly becomes visible. Once the band is complete, no spots remain visible.

Remarks: According to Pierre BRICHARD, the distribution at the former village of Bemba is restricted to no more than a kilometre of shoreline. The rocks are shared with *Tropheus* sp. "Bemba", the "Orange Band *Tropheus*" (Pop. 2.4). BRICHARD estimated the population back then as no more than a few thousand individuals. That is certainly not many. The littoral is gradually becoming silted by the heavy sediment burden of the Ruzizi, and it is foreseeable that in the course of time this population will become smaller and be wiped out in the not too distant future. This is also happening in other biotopes in the northern part of the lake. There is no longer any possibility of migrating in this area.

This photo is about 30 years old and shows a young adult fish. To the best of my knowledge, there are no recent reports on this colour variant of lineage 10.

photo: H. J. Richter, Archiv A.C.S.

Tropheus cf. *duboisi*; Population 10.2

Trade name: Tropheus duboisi "Yellow Band", Yellow Band Duboisi, Kigoma Duboisi

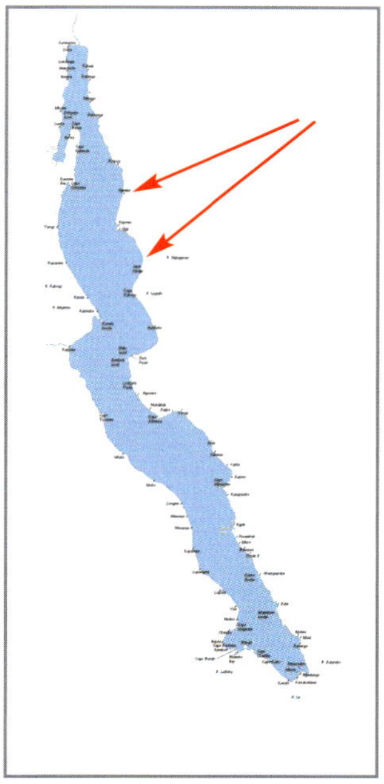

Distribution: This local form has the widest distribution within lineage 10. It extends south from Nyanza Lac to south of the Malagarasi delta.

Discovered: The existence of this colour form was discovered in several places in northern Tanzania by Wolfgang STAECK and others in the mid 1970s.

First importation: The first importations to Europe took place not long afterwards.

Coloration: The coloration differs from that of the population at Cape Munene in the northern Congo only in the width and colour of the band. The band is almost twice as wide as in individuals at Bemba and encompasses about 4-5 scales. A more or less yellow tint may be present in the upper part of the band. The density of the yellow colour varies from individual to individual. Juveniles exhibit the same colours and dots as individuals from other locations.

Remarks: In an aquarium with sufficient hiding-places, and given appropriate maintenance, these attractive fishes will rapidly breed. The male entices a ripe female to a suitable spot in the tank and spreads his anal fin on the substrate in front of his chosen one. The female mouths the fin, taking up sperm. After a while the pair change places and the female lays an egg directly in front of the mouth of the male. She quickly turns round and picks up the egg "guarded" by the male. And then the whole procedure is repeated. In full-grown the average spawn size is 8-12 eggs. The female must continue to be fed during the brooding period, as otherwise she will enjoy a meal of eggs. If the tank is not overpopulated and has a few peaceful spots then the initial release of the fry can be observed after four weeks at a water temperature of 27 °C. Brood care continues for several weeks, with the female taking her fry into her mouth when disturbed and at night, until the increasing size of the young makes this impossible. Given plenty of algae on the decor, freshly-hatched Artemia, and a little dried food the fry will thrive. If the young are left in the tank then their colour change can be observed when it begins a few months later. Such observations are not to be missed.

Tropheus cf. *duboisi*; Population 10.3

Trade name: Tropheus duboisi "Broad Band", Broad Band Duboisi, Tropheus duboisi "Maswa"

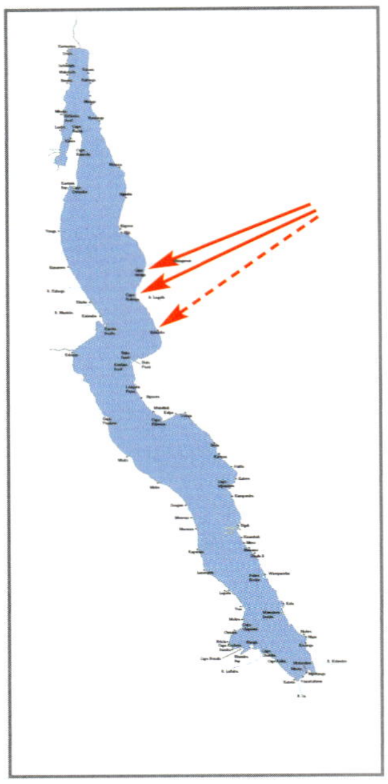

Distribution: The distribution of this colour form begins to the south of the Lugufu and extends at least as far as Cape Kabogo. The distribution may extend a few kilometres further south. Although available data are scant, it appears that the band is narrower in the southern part of the range (marked as Halembe on the distribution map of the lineage). These individuals have not been given a population number here.

Discovered: This local form was discovered by aquarists at the beginning of the 1980s.

First importation: This form was first imported for the ornamental fish industry shortly after its discovery. To the present day, this colour variant of *Tropheus* remains virtually a standard item in the relevant sector of the trade.

Coloration: The coloration is as in the preceding forms apart from the intensity of the colour and the width of the band. The anterior part of the head is, compared to the brilliant blue of the rest of the head, overlain with a rather whitish colour. Wild-caught fish may lose this anterior head colour in time. The yellow colour of the up to 7 scales wide band is more intense than in the other local forms. There are also individuals in which the band is narrower. The lower rows of scales in the band on the belly are always white. Dorsal and anal are slightly bluish, as is the base of the pectorals. In older juveniles the spots still present have a slight yellow tinge.

Remarks: There have been repeated reports (usually in April) of a red-banded *Tropheus duboisi*. The existence of such a fish is probably no more than the pipe-dream of the few. On the other hand, a few years ago I heard (from a normally reliable source) about a cross between *Tropheus duboisi* and the "Orange Band *Tropheus*" (2.4). These two *Tropheus* lineages became separate back in the distant mists of time. It would be worth experimenting to see whether such a mating is in fact possible, as a positive result would certainly raise new questions, and simultaneously provide some answers to some of the existing puzzles regarding the evolution of these fishes. Whether reality or pipe-dream, perhaps such hybrids are at the root of all the rumours. It would not surprise me if such a mating were to produce a *Tropheus* with a pale orange band. The existence of a *Tropheus duboisi* with a red or orange band cannot, however, be completely excluded, as the coast of Lake Tanganyika remains inadequately studied or totally unexplored in many places. We may yet be surprised.

A *Tropheus* cf. *duboisi* of population 10.3 in the transitional colour phase. photo: P. Schupke

Tropheus of population 10.4 live at the north-western tip of the Kungwe Mountains, where they were discovered by H. W. Dieckhoff at Karilani Island. **photo**: A. Konings

Tropheus cf. *duboisi*; Population 10.4

Trade name: Tropheus duboisi "Karilani", Karilani Duboisi.

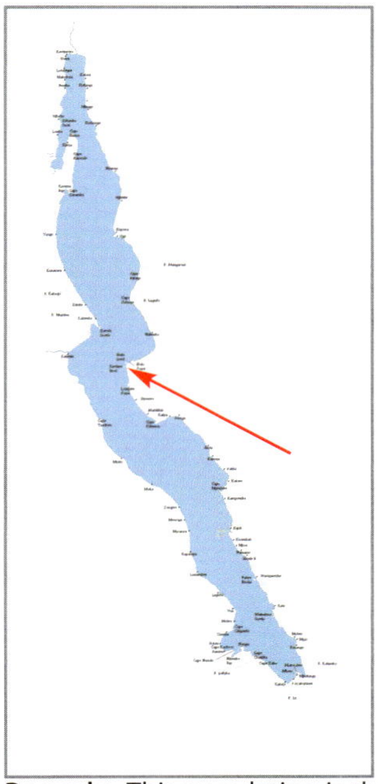

Distribution: As far as is known at present, the distribution of this local form lies at Magambo and close-by Karilani Island.

Discovered: This *Tropheus* was discovered by H. W. DIECKHOFF during the 1980s.

First importation: This local form was imported for the ornamental fish industry just a few times shortly after its discovery.

Coloration: The black colour of the body is interrupted by a very narrow white band that begins roughly beneath the 6th dorsal spine and extends to the belly. This white band is the narrowest in all the *Tropheus* of lineage 10, and is often only 2-3 scales wide. The brilliant blue colour of the head is especially prominent. The dorsal and anal fins exhibit the same shade, but are not as intensely coloured. The pectorals have the bluish colour of the head only at their base. Juveniles exhibit the usual pattern of dots and are indistinguishable from the juveniles of other populations of lineage 10.

Remarks: This population is the most recently discovered colour form of lineage 10, and was not discovered until the 1980s. The discovery of a new population of this species so far south created a minor sensation at the time. *Tropheus duboisi* is found sympatric with 5 other *Tropheus* species at Karilani Island. However, each of the 5 species have different food and territory requirements, as otherwise they would not be able to live together. They live at different depths and utilise different elements of the local Aufwuchs, so that they are not in competition. The rocky coast continues south with no notable interruptions, and so it is within the realms of possibility that additional *Tropheus duboisi* populations may be discovered further to the south.

Tropheus of lineage 11

The *Tropheus* from the area north of Ikola are without doubt among the best known and most colourful members of the genus. Their bright lemon-yellow band, which is little affected by mood, has made them highly popular with enthusiasts. When the first specimens appeared in the trade in the mid 1970s, they were thought to be close relatives of the black *Tropheus* in the north, in this book classified as lineage 1. The colours and their distribution on the body make such a relationship seem probable at first glance. But doubt must be cast on this theory in view of their location, which lies far from the localities for *Tropheus* of lineage 1. Research by STURMBAUER and MEYER has confirmed this suspicion. The black *Tropheus* from the north separated very early, geologically speaking, from one of the main lineages existing at that time. The *Tropheus* at Ikola are thought to have done likewise. The "Kaiser Moorii" is now assigned to a lineage of its own. It bears a certain external resemblance to lineage 5, with which it shares various mood-related colour patterns. Only the colour of the iris betrays the fact that the fishes in question belong to different lineages. The origins of the Ikola *Tropheus* lie in one of the main lineages which produced the various lineages that compose lineage 3. Although the *Tropheus* from Ikola are among the most frequently collected species of the genus, both in the past and today, little is known regarding their distribution. Only a few enthusiasts have made the effort to travel to the area and investigate the limits of the distribution of these fishes. A group of Swedish aquarists stayed in the area for a short time. Further to the north of Ikola they individuals with reddish bases to the scales in the yellow band. Nothing further is known. Pierre Brichard mentions in his book on Tanganyika cichlids that the fishes he had seen had orange flanks and a belly of the same colour, and the most beautiful specimens even had an orange head and anterior body. From this quotation it is quite clear that BRICHARD must have seen different *Tropheus* "Kaiser" to those with which we are generally familiar. There have been no reports of other *Tropheus* species that might perhaps inhabit other parts of the littoral or at least coincide with the Ikola *Tropheus* at the limits of its range. Only recently have scientists from University of Innsbruck managed to gather information on the spot. It transpires that to the north the rocky littoral drops off to ever greater depths. Here they found *Tropheus polli* within the distribution of the Kaiser Moorii, forcing the Kaiser into deeper water. It is conceivable that in deeper water the yellow colour of *Tropheus* sp. "Ikola" changes to more of an orange, as described by BRICHARD. But such individuals have not been found. The extent and intensity of the coloration can vary considerably from individual to individual. Aquarists are generally familiar only with the appearance of wild-caught specimens. Those enthusiasts that have managed to breed the Ikola *Tropheus*, are also aware of the differences that may occur in the width and colour of the band (see photos). The collectors on the spot catch the individuals with the most regular and brightest band, and it is only yin their offspring that we see the huge range of possibilities present in the gene pool. *Tropheus* sp. "Ikola" has five anal spines and an almost straight-edged tail. The upper part of the iris is red and the remainder dark, or silvery with a bluish sheen, depending on mood.

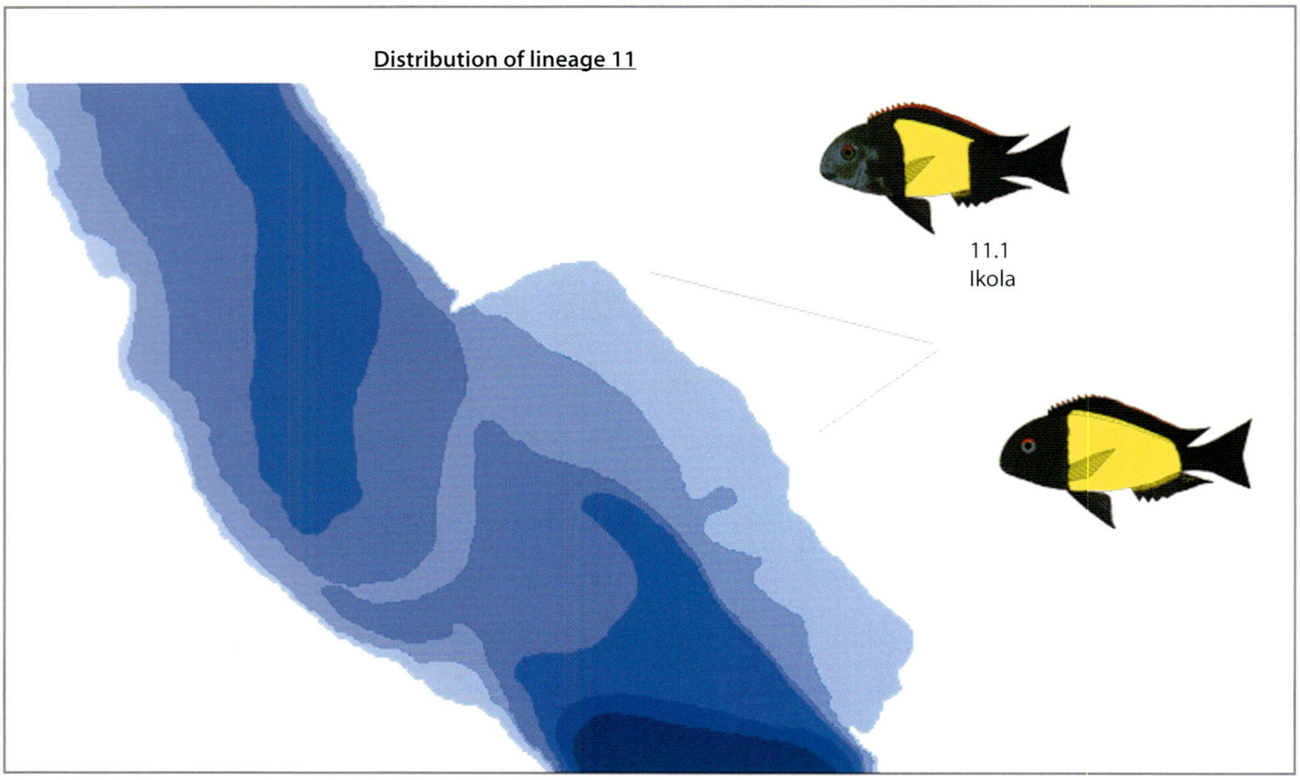

Tropheus sp. Ikola; Population 11.1

Trade name: Tropheus sp. "Ikola", Kaiser Moorii.

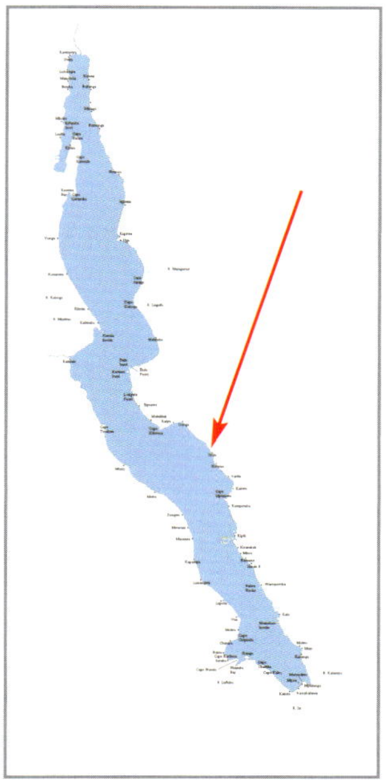

Distribution: The distribution begins to the north of the village of Ikola in Tanzania. The northern limit of the range is a few kilometres south of the village of Isonga.

Discovered: This *Tropheus* was discovered by FAINZILBER in the mid 1970s..

First importation: The first importation of small numbers took place only a short time afterwards. Not until a few years later was this form imported regularly in large numbers. There have often been breaks of several years between consignments.

Coloration: In general the anterior part between the mouth and the 3rd or 4th dorsal spine and the ventral insertion is black-brown. Thereafter there is a wide lemon-yellow band that extends to the soft parts of the dorsal and anal. The yellow band may be variable from individual to individual; thus there are specimens in which the band does not begin until a few scales further back, and others in which the upper part of the band ends earlier so that its posterior edge is slanting rather than straight. The lower half of the dorsal may be yellow, but there are also individuals in which only a narrow area at the base of the dorsal is yellow. There are specimens, albeit rare, in which the band is divided in two. Vertically (ie one vertical band in two parts, top and bottom) or horizontally (ie two vertical bands. There are also individuals in which the band extends to the caudal peduncle and penetrates to the middle of the caudal, colouring it yellow, so that only the posterior edge remains black. The iris in lightened individuals is red in its upper part and silvery with a slight bluish sheen in the remaining, lower, part. According to Wolfgang STANGELMAYER, the band is rather broader and lighter north of Ikola than further north at Kekese, where the fishes additionally have a red marginal band in the dorsal.

Remarks: Variations in the width of the band are seen more often in tank-breds. These extreme variations are seen mainly in tank-breds because wild fish are collected selectively. Depending on mood, the anterior dark area may lighten to a greenish yellow-grey, seen in spawning females and under stress. The fish in the left-hand photo below has become much lighter due to stress and its band has decreased considerably in width. An almost identical mood-related coloration can be seen in the *Tropheus* sp. "Karago", population 5.10, in the photo below right. The two fishes can be distinguished only by eye colour. This similarity must not be taken to indicate a close relationship, but may point to a shared ancestor. *Tropheus* sp. "Ikola" crosses successfully with *Tropheus* sp. "Mabilibili" (Pop. 12.1, qv).

In fright coloration *Tropheus* sp. "Ikola" exhibits a confusing resemblance to *Tropheus* sp. "Karago" from lineage 5, apart from the differently coloured iris.

And vice versa: *Tropheus* sp. "Karago" in fright coloration looks very similar to a *Tropheus* sp. "Ikola", apart from the differently coloured iris.

photos: P. Schupke

Along with the *Tropheus* "Kaiser", *Tropheus* "Kirschfleck" is among the most popular and desirable of all *Tropheus*. They were known to aquarists as early as the mid 1970s. The two red spots on a black background are very striking, at least in the aquarium, and are little affected by mood. The spots are almost always visible. Only when aroused do females and males (whose head remains black) lighten all over to pale grey, apart from the tail and the red spots. The previous bold contrast is then appreciably reduced.

According to details from importers, in some places at least *Tropheus* "Kirschfleck" prefers the deeper rocky littoral at about 7-10 m, while the upper habitat is the home of brownish *Tropheus*, which may belong to the Karilani colour form assigned to lineage 3. At this depth the red is no longer recognisable as that colour. Moreover the light grey courtship dress with red spots and black caudal and head looks different at this depth compared to in our aquaria.

The approximately 70 kilometre distribution, about 50 km of which form part of the Mahale National Park, contains several colour populations. The colour spectrum extends from the cherry red so popular with aquarists to orange-red, yellow with a red reticulation, matt yellow, and ivory coloured with a yellow background and red bases to the scales. In the south of its range the "Kirschfleck" encounters *T.* sp. "Kibwesa", which belongs to lineage 6, and whose home is the central part of the lake, rarely visited by aquarists. In the north lineage 12 is sympatric with three different lineages (3, 4, and 10). This area was the source of the early imports that reached us at long intervals. The fishes with the blood-red spots originate from the south, in the vicinity of Mabilibili. According to recent research, in the past, during fluctuations in the lake level, *Tropheus* "Kirschfleck" came together with other *Tropheus* lineages, and its genetic material is correspondingly heterogenous. Thus migrant *Tropheus*, close to the current lineage 3, left traces in the genes of the "Kirschfleck". The early ancestor of this now separate lineage apparently settled in the central lake and, over the course of the ages, evolved into the species we see today. It is quite conceivable that *Tropheus* of lineage 3 migrated into the central basin when the three lakes became united and forced the long-resident *Tropheus* into deeper water in some places. The *Tropheus* of lineage 5 between the Kavala Islands and Kunanwa further to the north apparently underwent something similar. In their case the population level decreased, perhaps because they were unable to move to deeper water. Perhaps one day detailed on-the-spot research will throw a little more light on the evolutionary history of *Tropheus* and the consequences of the repeated fluctuations in the level of Lake Tanganyika in the past. We are certainly not jumping the gun in assuming that ancient and younger lineages are in competition for the best habitats in the lake today.

The central part of the lake was for a long time isolated from the other basins, and lineage 4, to which *Tropheus polli* belongs, would have been able to colonise the entire shoreline of the central lake at that time. Something similar may have happened with the lineage that is today represented by *Tropheus* "Kirschfleck". There are *Tropheus* of lineage 3 on the west coast as well, and if these did not migrate there until a later date, they may perhaps have suppressed a lineage already resident there. Thus the hope is quite justified that *Tropheus* of this ancient lineage, today represented on the east coast by *Tropheus* "Kirschfleck", may have survived in deeper water in a few places. The area was in fact privately explored some years ago, but the information obtained was not publicised as widely as might have been desired. The latest information on lineage 12 comes from Ad Konings, who has discovered a black *Tropheus*, which he considers close to the "Kirschfleck" at Lubugwe, a few kilometres south of Luagala Point (Information from the Internet).

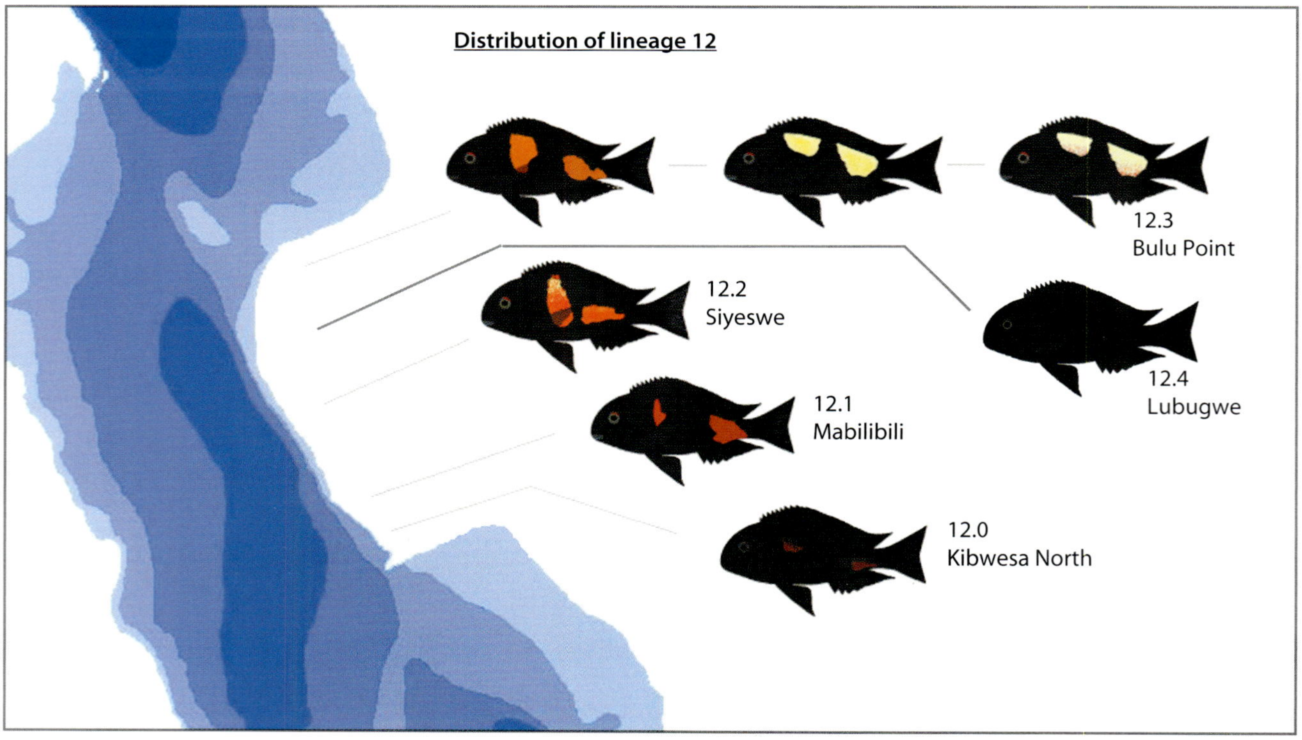

Tropheus sp. Kibwesa Nord; Population 12.0; *Tropheus* sp. Mabilibili; Population 12.1

Trade name: Tropheus "Kirschfleck", Tropheus "Cherry Spot", Kirschfleck Moorii

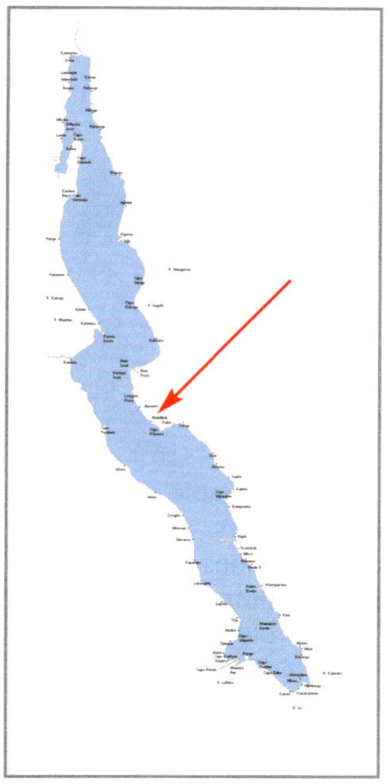

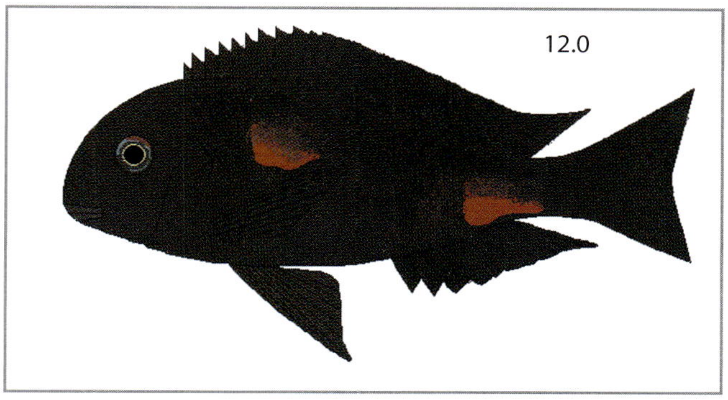

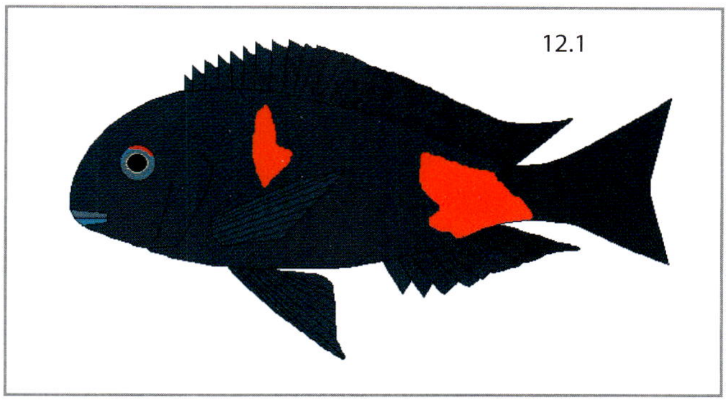

Distribution: The locality lies at the southernmost point of the distribution of this lineage, in the Mahale National Park.

Discovered: The colour form 12.1 was discovered a few years after the form from Bulu Point, which Fainzilber discovered in 1975.

First importation: The first importation probably took place at the beginning of the 1980s.

Coloration: The base colour of these *Tropheus* is black, which may lighten to a pale grey depending on mood. The caudal and its peduncle, however, remain permanently black. On the body there are two large blood-red spots, individually very variable in their extent, which as the name "Cherry Spot" suggests, look very much like two cherries. When the fish is aroused its appearance alters considerably. During courtship females change to a light grey. The red spots and the posterior black region (the caudal and its peduncle) remain unchanged. Subordinate males may exhibit the same coloration. Sexually active males also retain the black anterior area, with only the part containing the spots changing to pale grey. In spawning females and subordinate males the red colour may darken to some extent. All the fns are black. Areas where the coloured spots contact the dorsal may become lighter. In many individuals the caudal exhibits a light grey, somewhat translucent triangle whose apex points towards the caudal peduncle. The upper edge of the eye is the same red colour as the spots, while the rest of the iris has a rather bluish cast. The most southerly individuals (those from north of Kibwesa) are very dark in colour; the spots are whitish with a rather low or non-existent red component. This form is here numbered 12.0, although transitional forms are to be expected.

Remarks: *Tropheus* "Kirschfleck" and the "Kaiser Moorii" should not be kept together as they can successfully hybridise. This even happened to a collecting team who, after visiting the collecting site for *Tropheus* "Kaiser", had to keep the latter for some time in large nets in Kirschfleck *Tropheus* territory while they collected in the latter. The next day the nets were torn and the Kaisers had found themselves a new home. Since this accident there have often been hybrids in Kirschfleck consignments. These fishes are attractively coloured. In order to make a virtue of necessity, records should be made of how these mixed populations get on under the local conditions and the long-term effects on the local population of *Tropheus* "Kirschfleck". What happened as an accident in this case, and, of course, cannot be undone, has also occasionally happened in Zambia through carelessness.

Hybrids between "Kaiser" and "Kirschfleck" are attractive in appearance. But they should be recognised for what they are.
photo: P. Schupke

Tropheus sp. Siyeswe; Population 12.2

Trade name: Tropheus "Kirschfleck", Tropheus "Cherry Spot", Kirschfleck Moorii

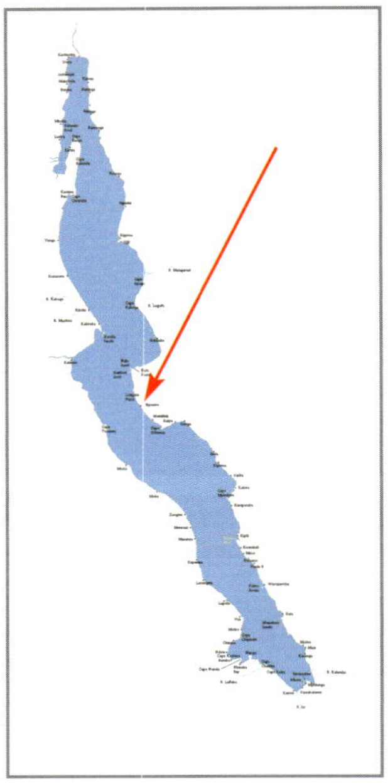

Distribution: The distribution of this colour form is said to lie in the general area of the village of Siyeswe.

Discovered : This colour form was discovered in the mid 1980s.

First importation: The first importation took place in the same year.

Coloration: The almost black base colour is broken by two light red spots of variable size. There are individuals in which the spots are yellow-tinged, and then appear lighter. The most attractive specimens, however, have plain, brilliant light red spots. The fin colour matches the dark body, though areas where the spots border the bases of the fins may lighten somewhat to greyish. Sometimes the dorsal and anal appear somewhat lightened. The slightly bluish iris and its red upper edge are difficult to discern in darkened individuals. Very young individuals have poorly delineated, whitish vertical stripes, but these rapidly disappear. The sites of the future spots can be detected at a very early stage. Red coloration can be seen in the dorsal and the gill-cover area in juveniles, but this disappears in the course of time.

Remarks: Sometimes tank-bred specimens occur in which the spots merge and virtually form a broad coloured band. Such aberrant forms can be mated with similar specimens in order to produce broad-banded individuals. This process requires a degree of patience, as it may extend over several generations, ie several years. The results are certainly worth seeing, as the colour combination is almost unsurpassable. What has long been practised in other areas of the aquarium hobby, ie the development of cultivated forms, is virtually a taboo subject in the field of Lake Tanganyika cichlids. But the first steps have already been taken. In Zambia yellow-orange *Tropheus* sp. "Kalambo" have been deliberately mated to one another in order to stabilise their aberrant genes. Because, fortunately, tastes differ, completely yellow *Tropheus* sp. "Kalambo" and perhaps *Tropheus* "Kirschfleck" with a complete, broad red-orange band, will find their fans, but they should then be given new names. In 2001 a brown *Tropheus* was discovered at an exporter´s, which displayed a broad red-orange band during courtship. The collecting site could not be determined.

Tropheus sp. Bulu Point; Population 12.3

Trade name: Tropheus "Kirschfleck", Tropheus "Cherry Spot", Kirschfleck Moorii

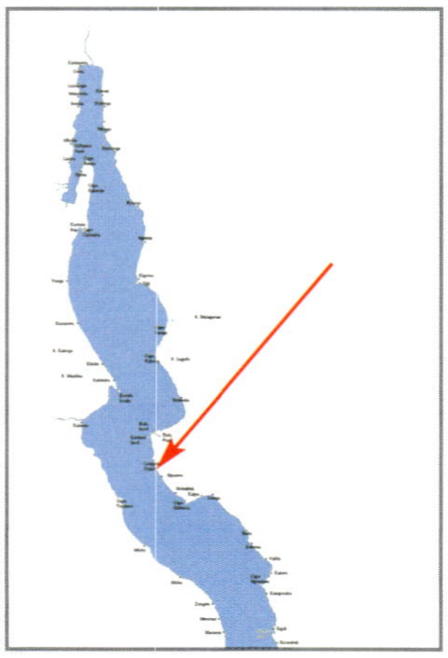

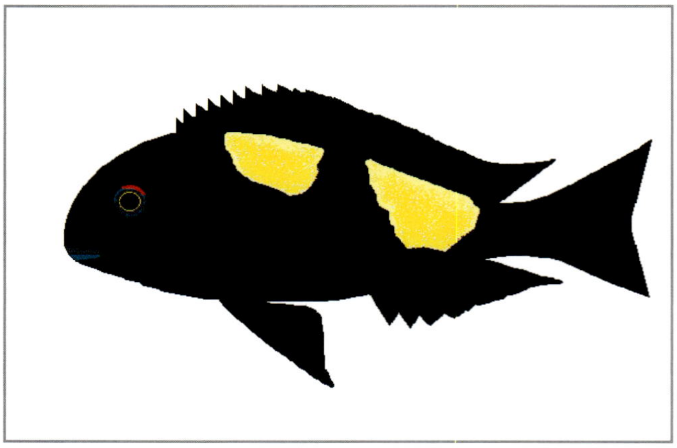

Distribution: This colour form is said to occur in the general area of Bulu Point. Information regarding the provenance of this fish is often contradictory.

Discovered: This colour form was discovered by Fainzilber in 1975 and documented by Wolfgang Staeck in 1977.

First importation: These *Tropheus* were formerly often imported, but later on fell out of favour with aquarists on account of their not so vivid colours.

Coloration: The base colour is a dark black-brown, with two orange, orange-yellowish, or yellowish white spots on the body; these may be overlain with an orange reticulation. Individuals with orange spots are purportedly the exception. The size of the spots and their position on the body are different for every fish and can be used to identify the individual. The anterior spot may be reduced to a narrow streak, but equally may occupy a very large area and be much bigger than the posterior spot. The same applies to the posterior spot. Where the spots contact the dorsal and anal, the base of the fin(s) is coloured to some degree and a small part of the fin is much lighter. The lower half of the body may have a faint bronze sheen. The fins are normally all the same colour as the body, but it is possible that when a spot lies close to a fin or actually adjoins it, the corresponding part of the fin may be somewhat lighter. This applies to the dorsal and anal. The caudal may have a small, slightly translucent, light central triangle posteriorly. The slightly bluish iris has red on its upper outer edge. Some individuals are plain-coloured, without spots.

Remarks: This colour form is not all that popular with aquarists, but was in the past sometimes a favourite with collecting teams as the locality could be reached very rapidly during their collecting itineraries. In those days the route ran from Kalemie, out to the Kavala Islands, and thence straight across to the Tanzanian coast. The collecting schedule was arranged so as to reach the Tanzanian coast on Friday, as most of the people there are Moslems and Friday is their day of rest. A favourite trick was to later on mix these *Tropheus* with more attractive forms caught elsewhere, but this was unacceptable to aquarists, and ultimately importers. In order to meet the stringent requirements of the customer, it was necessary to go further south to catch bright red coloured fishes, which meant taking longer travelling times into account. Within their range the *Tropheus* of lineage 12 are said to be outnumbered by *Tropheus* of lineage 3.

Tropheus sp. Kabezi, Linie 13, Population 13.0

Trade name: None known.

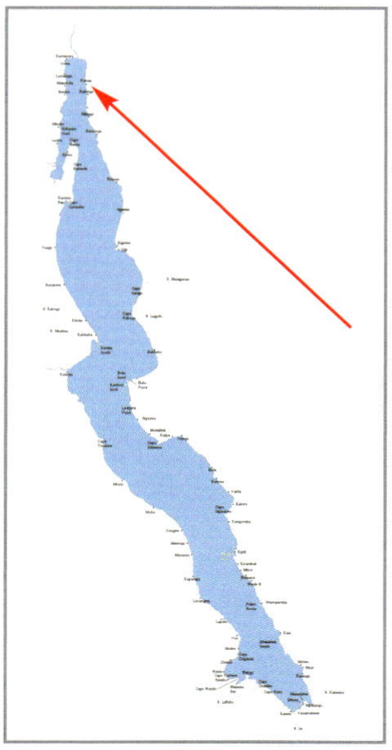

This population was only recently discovered in the vicinity of the village of Kabezi, a few kilometres south of Bujumbura. The fishes are completely black in colour. Gene sequencing by zoologists from the University of Innsbruck produced the remarkable discovery that these *Tropheus* derive from a very ancient lineage. They belong to the distant relatives of lineage 3, which spread its genes all over the lake at an early stage. Its ancestors were able to become more widespread as a result of fluctuations in lake level, and, as already mentioned, spread their genes right down to the southern end of the lake. It is quite conceivable that *Tropheus* sp. "Kabezi" was formerly distributed further to the north before the Ruzizi came into existence.

In 1999, 2000, and 2001 new *Tropheus* populations were discovered at various sites around Lake Tanganyika. The locality data cannot be confirmed and must accordingly be treated with appropriate caution.

In 1999 a population from Bulu Point reached whose colour pattern does not permit it to be readily assigned to any existing lineage. *Tropheus* "Red Belly", as it is known, exhibits certain similarities to the fishes of lineage 9 between Karema and Namansi, but in fact any relationship can be ruled out because of the great distance from lineage 9. There is also a similarity in colour pattern to that of population 3.6 from the general area of Kalemie, which, as the population number indicates, belongs to lineage 3. Gene sequencing may perhaps provide some clarification. According to data from the collector who caught these fish at the time, this *Tropheus* population is very rarely encountered. It was caught in the same area as *Tropheus* sp. "Kirschfleck" (pop. 12.3). In December 2001 *Tropheus* "Red Belly" was imported again, and I hope that I will soon be able to provide more information and photos.

In summer 2001 *Tropheus* "Karilani Red", as the collectors named it, was captured n the immediate vicinity. Apart from the red bar on the caudal peduncle, the coloration of older individuals resembles that of *Tropheus* sp. "Chipimbi" (pop. 7.5). Younger specimens, by contrast, are appreciably lighter and have a rather indistinct, faint reddish area on the body (hence, perhaps, the name) and in the dorsal. A second importation of the "Karilani Red" shows a slight colour variation. These fishes have an attractive bronze sheen on the body. The blue eye-ring has bright red in its upper part. While making the final corrections to this book (in July 2002) the author found a new *Tropheus* on the Internet, labelled *Tropheus* sp. "Red Sunspot" and supposedly caught in Zambia. At Karilani Island there are supposedly also *Tropheus* sp. "Kirschfleck", *Tropheus polli* (or *T. annectens*), and *Tropheus duboisi*. The familiar *Tropheus* sp. "Karilani", with its striking yellow pectorals, is not found until further north, in contradiction to the data given by KONINGS. Because of the absence of information, *Tropheus* sp. "Karilani Red" cannot be assigned to any lineage. Here too DNA research may provide additional clarification.

A further hitherto unknown *Tropheus* population was discovered not in the lake, but in an exporter´s holding tank in Dar Es Salaam, Tanzania. Unfortunately there was only a single specimen, but the importer brought it back to Germany and kindly allowed me to make additional observations. Its less than inviting brown colour alters when it is active and displaying, so that the fish looks quite attractive. Apart from the anterior third, which remains brown, there is a light grey area, overlain with a heavy orange reticulation, extending to the caudal peduncle and creating the impression of an orange-yellow band. When the fish is most highly motivated the bases of the scales in this band are yellow, but otherwise more light grey. The back immediately below the dorsal is light grey. The grey colour can extend for up to half the width of the band in the form of vertical bars. Dorsal, ventrals, and caudal are a light grey-blue, similar to the colour on the upper dorsum. The locality for this form remains unknown.

The other new discoveries are not difficult to assign to existing lineages on the basis of their appearance. Recently more detailed investigation of the locality for *Tropheus* sp. "Kabwe" (pop. 9.7), discovered back in 1994, revealed individuals with aberrant pigmentation. In autumn 2001 I was able to marvel at a number of these bizarre *Tropheus* at the "Cichlid Shack" of Helmut LÖFFLAD. The dark brown of the regular coloration is broken by spots of variable size where melanin is completely or partially lacking. The smaller the amount of dark pigment, the brighter the yellow of the affected area. It may be that such individuals are no less well camouflaged than normal-coloured conspecifics in the constantly varying play of the sunlight in the upper rocky littoral. See pp. 113-114.

It was only a few years ago that we became aware of *Tropheus* sp. "Namansi" of lineage 9, whose juveniles have a bright red component (see colour forms of lineage 9). The newly-discovered aberrantly-pigmented variant of this population should be mentioned here. The relatively light (at least in the aquarium) normal form is somewhat surpassed by the mutated form. The home page (well worth a visit) of the company African Diving includes a photo of an almost completely yellow-orange *Tropheus*. Only the anterior dorsal contains an area corresponding to the colour of the normal form. Because only this one specimen is pictured and I know of no other photos, it is impossible to say whether or not the lightening of the body varies with the individual. I know of no importations of this *Tropheus* to Germany.

South of Lunangwa in the Congo a new population has been discovered that belongs to lineage 7 on the basis of its appearance. Unfortunately only a single photo is available, on the Internet. The photo quality is adequate to permit a brief description of this population. The fins and head are very dark, as is the area immediately below the dorsal, which extends down onto the dark red area of the flank in the form of the beginnings of vertical bars. As in the other members of lineage 7, the appearance will be dependent on different moods. I do not know whether these *Tropheus* are available in Germany yet.

A very attractively coloured form reached us almost unnoticed in 1999, and we (at least those of us in southern Germany) became aware of it only two years later. In heavily polluted water these *Tropheus* are inclined to show no colour. The dark wine-red head contrasts boldly with the light yellow lower body. The upper third of the body has a greenish-yellow sheen. Dorsal, anal, and pectorals are yellow, and the caudal peduncle almost white. This population is probably a colour variant of the Yellow Rainbow *Tropheus*, which belongs to lineage 8. It cannot be ruled out that it may perhaps be a light-coloured form of *Tropheus* sp. "Ilangi". According to Pierre BRICHARD, in the southern entrance to Nkamba Bay there are two populations belonging to different lineages, which do not, however, mingle. An assertion that to date has not been more closely investigated.

Neuentdeckungen / Recent discoveries

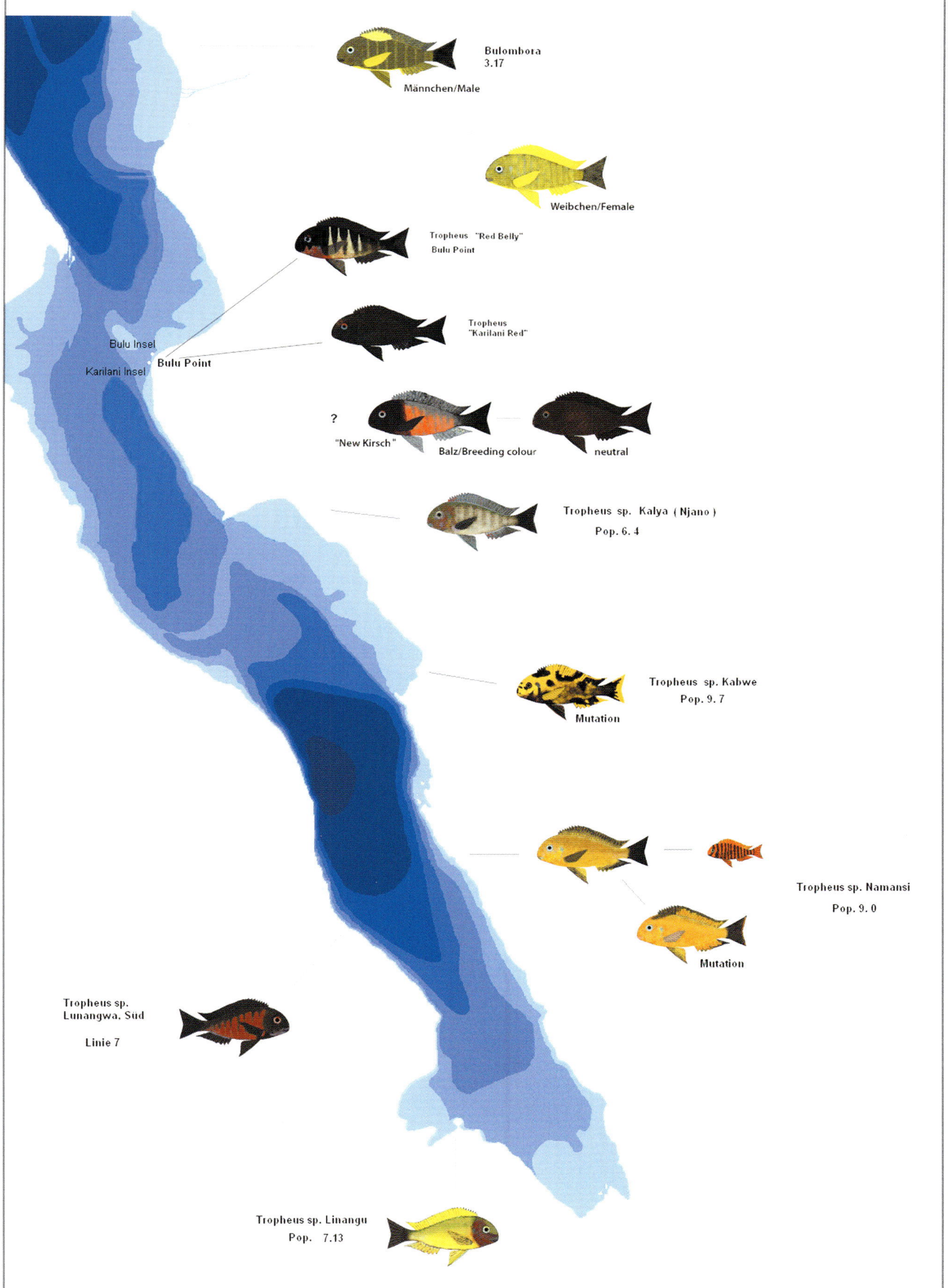

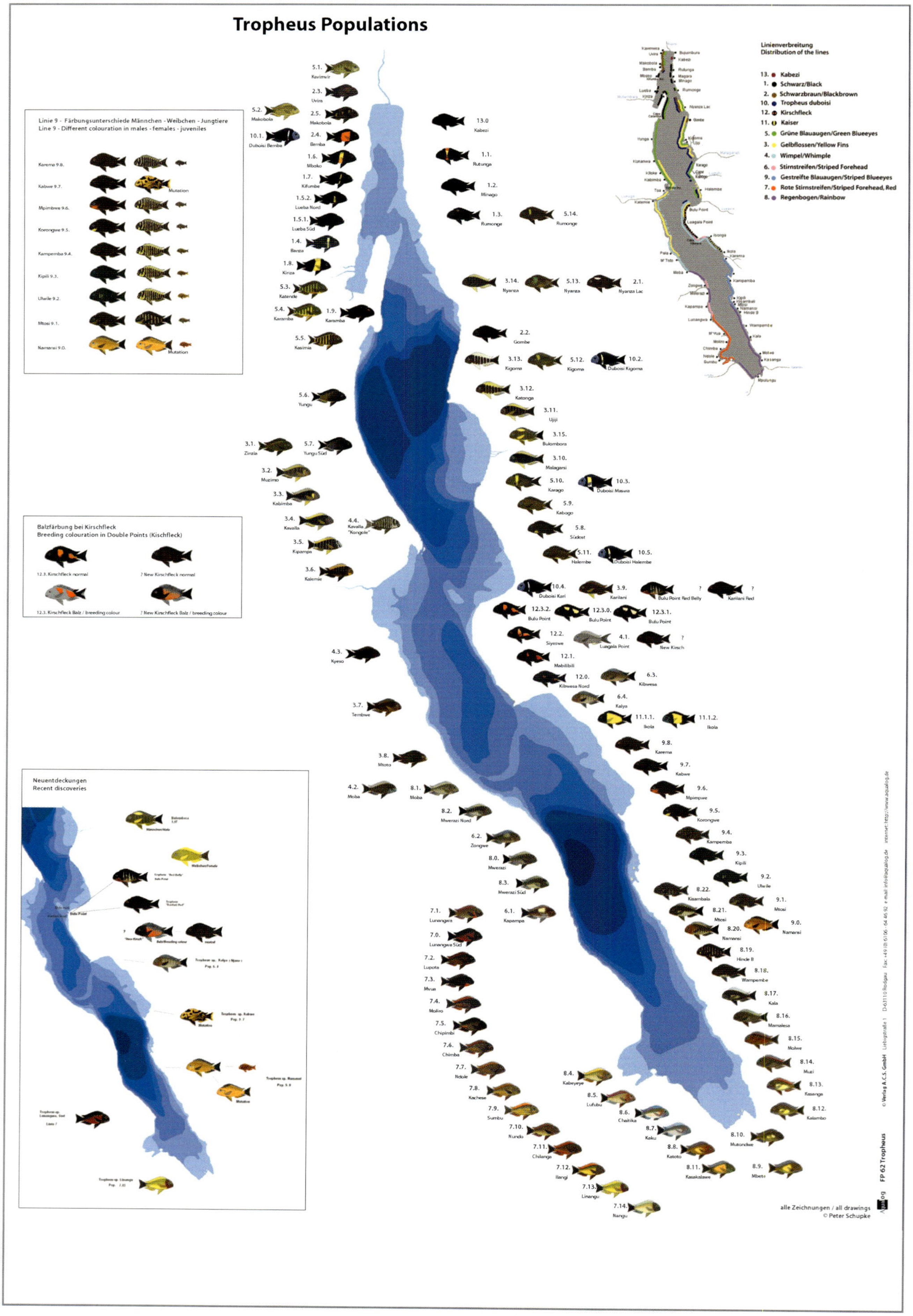
Tropheus Populations
Linienverbreitung
Distribution of the lines
13. Kabezi
1. Schwarz/Black
2. Schwarzbraun/Blackbrown
10. Tropheus duboisi
12. Kirschfleck
11. Kaiser
5. Grüne Blauaugen/Green Blueeyes
3. Gelbflossen/Yellow Fins
4. Wimpel/Whimple
6. Stirnstreifen/Striped Forehead
9. Gestreifte Blauaugen/Striped Blueeyes
7. Rote Stirnstreifen/Striped Forehead, Red
8. Regenbogen/Rainbow
Linie 9 - Färbungsunterschiede Männchen - Weibchen - Jungtiere
Line 9 - Different colouration in males - females - juveniles
Karema 9.8.
Kabwe 9.7.
Mutation
Mpimbwe 9.6.
Korongwe 9.5.
Kampemba 9.4.
Kipili 9.3.
Ulwile 9.2.
Mtosi 9.1.
Namansi 9.0.
Mutation
Balzfärbung bei Kirschfleck
Breeding colouration in Double Points (Kischfleck)
Neuentdeckungen
Recent discoveries
5.1. Kavimvir
2.3. Uvira
5.2. Makobola
2.5. Makobola
10.1. Duboisi Bemba
2.4. Bemba
1.6. Mboko
1.7. Kifumbe
1.5.2. Lueba Nord
1.5.1. Lueba Süd
1.4. Baraka
1.8. Kiriza
5.3. Katende
5.4. Karamba
1.9. Karamba
5.5. Kasimia
13.0 Kabezi
1.1. Rutunga
1.2. Minago
1.3. Rumonge
5.14. Rumonge
3.14. Nyanza
5.13. Nyanza
2.1. Nyanza Lac
2.2. Gombe
3.13. Kigoma
5.12. Kigoma
10.2. Duboisi Kigoma
3.12. Katonga
3.11. Ujiji
3.15. Bulombora
3.10. Malagarasi
5.10. Karago
10.3. Duboisi Maswa
5.9. Kabogo
5.8. Südost
5.11. Halembe
10.5. Duboisi Halembe
5.6. Yungu
3.1. Zinzia
5.7. Yungu Süd
3.2. Muzimo
3.3. Kabimba
3.4. Kavalla
4.4. Kavalla "Kongole"
3.5. Kipampa
3.6. Kalemie
10.4. Duboisi Kari
3.9. Karilani
Bulu Point Red Belly
Karilani Red
12.3.2. Bulu Point
12.3.0. Bulu Point
12.3.1. Bulu Point
12.2. Siyeswe
4.1. Lugala Point
New Kirsch
12.1. Mabilibili
12.0. Kibwesa Nord
6.3. Kibwesa
6.4. Kalya
11.1.1. Ikola
11.1.2. Ikola
4.3. Kyeso
3.7. Tembwe
3.8. Mtoto
4.2. Moba
8.1. Moba
8.2. Mwerazi Nord
6.2. Zongwe
8.0. Mwerazi
8.3. Mwerazi Süd
9.8. Karema
9.7. Kabwe
9.6. Mpimpwe
9.5. Korongwe
9.4. Kampemba
9.3. Kipili
9.2. Ulwile
9.1. Mtosi
9.0. Namansi
8.22. Kisambala
8.21. Mtosi
8.20. Namansi
8.19. Hinde B
8.18. Wampembe
8.17. Kala
8.16. Mamalesa
8.15. Molwe
8.14. Muzi
8.13. Kasanga
8.12. Kalambo
8.10. Mutondwe
8.9. Mbete
8.11. Kasakalawe
8.8. Katoto
8.7. Kaku
8.6. Chaitika
8.5. Lufubu
8.4. Kabeyeye
6.1. Kapampa
7.1. Lunangwa
7.0. Lunangwa Süd
7.2. Lupota
7.3. Mvua
7.4. Moliro
7.5. Chipimbi
7.6. Chimba
7.7. Ndole
7.8. Kachese
7.9. Sumbu
7.10. Nundo
7.11. Chilanga
7.12. Ilangi
7.13. Linangu
7.14. Nangu
alle Zeichnungen / all drawings
© Peter Schupke
FP 62 Tropheus
© Verlag A.C.S. GmbH

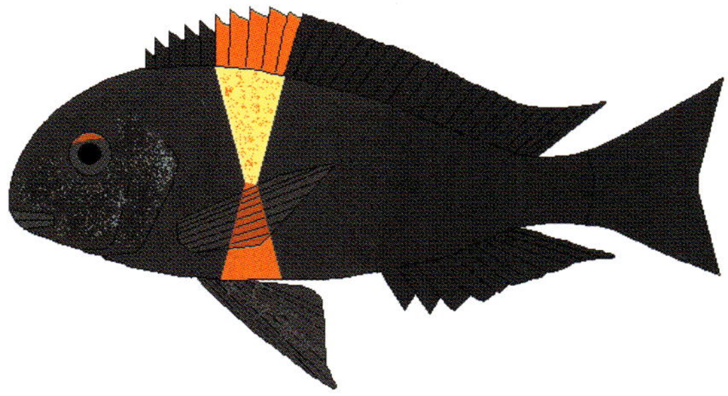

A97267 *Tropheus* sp. Rutunga; Population 1.1
Tropheus "Brabant", Tropheus "Rutunga"
From Kabezi to Magara, D.R. Congo, W, B, 16 cm

drawing: P. Schupke

A97267-5 *Tropheus* sp. Rutunga; Population 1.1
Tropheus "Brabant", Tropheus "Rutunga"
From Kabezi to Magara, D.R. Congo, W, B, 16 cm

photo: P. Schupke

A97267-3 *Tropheus* sp. Rutunga; Population 1.1
Tropheus "Brabant", Tropheus "Rutunga"
From Kabezi to Magara, D.R. Congo, W, B, 16 cm

photo: P. Schupke

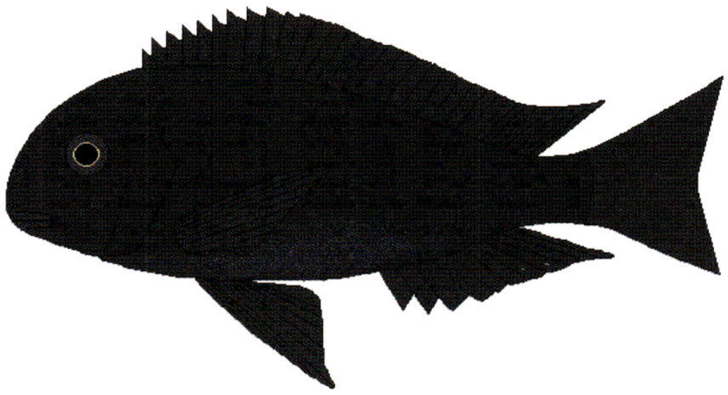

A97268 *Tropheus* sp. Minago; Population 1.2
Tropheus "Minago", Tropheus "Black"
South of the village of Minago, Burundi, W, B, 16 cm

drawing: P. Schupke

A97268-4 *Tropheus* sp. Minago; Population 1.2
Tropheus "Minago", Tropheus "Black"
South of the village of Minago, Burundi, W, B, 16 cm

photo: P. Schupke

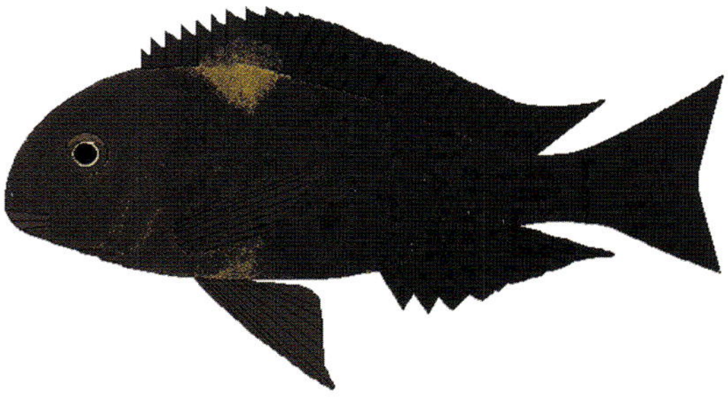

A97269 *Tropheus* sp. Rumonge I; Population 1.3

North of the village of Rumonge, Burundi, W, B, 16 cm

drawing: P. Schupke

A97270 *Tropeus* sp. Banza; Population 1.4

Manga to Cape Mizimo, D. R. Congo, W, 16 cm

drawing: P. Schupke

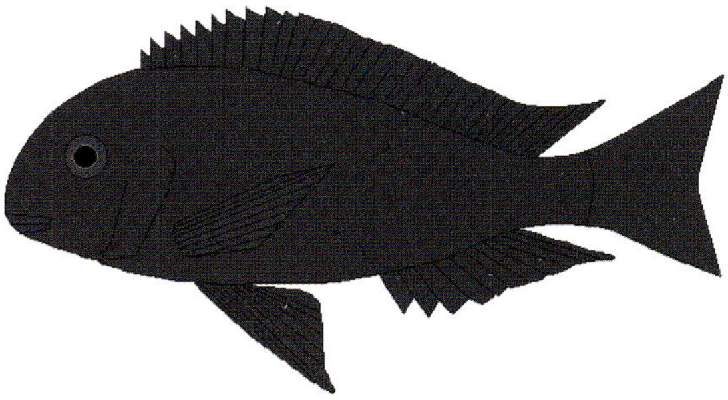

A97271 *Tropheus* sp. Lueba; Population 1.51

Lueba, D. R. Congo, W, 16 cm

drawing: P. Schupke

A97272 *Tropheus* sp. Lueba; Population 1.52

North of the village of Lueba, D. R. Congo, W, 16 cm

drawing: P. Schupke

A97273 *Tropheus* sp. Mboko; Population 1.6
Orange-Moorii I
North of the village of Mboko, D. R. Congo, W, B, 16 cm

drawing: P. Schupke

A97273-3 *Tropheus* sp. Mboko; Population 1.6
Orange-Moorii I
North of the village of Mboko, D. R. Congo, W, B, 16 cm

photo: P. Schupke

A97273-3 *Tropheus* sp. Mboko; Population 1.6
Orange-Moorii I
North of the village of Mboko, D. R. Congo, W, B, 16 cm

photo: P. Schupke

A97273-3 *Tropheus* sp. Mboko; Population 1.6
Orange-Moorii I
North of the village of Mboko, D. R. Congo, W, B, 16 cm

photo: P. Schupke

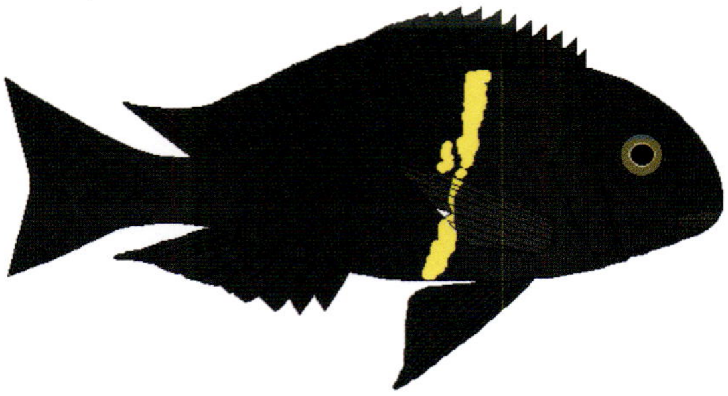

A97274 *Tropheus* sp. Kifumbwe; Population 1.7
Tropheus "Lemon Stripe"
Island near Mboko, D.R. Congo, W, B, 16 cm

drawing: P. Schupke

A97274-4 *Tropheus* sp. Kifumbwe; Population 1.7
Tropheus "Lemon Stripe"
Island near Mboko, D.R. Congo, W, B, 16 cm

photo: W. Staeck

A97275 *Tropheus* sp. Kiriza; Population 1.8
Tropheus "Kiriza", Yellow Kaiser
Near Uwari peninsula, D. R. Congo, W, B, 16 cm

drawing: P. Schupke

A97275-4 *Tropheus* sp. Kiriza; Population 1.8
Tropheus "Kiriza", Yellow Kaiser
Near Uwari peninsula, D. R. Congo, W, B, 16 cm
photo: P. Schupke

A97275-4 *Tropheus* sp. Kiriza; Population 1.8
Tropheus "Kiriza", Yellow Kaiser
Near Uwari peninsula, D. R. Congo, W, B, 16 cm
photo: P. Schupke

A97275-4 *Tropheus* sp. Kiriza; Population 1.8
Tropheus "Kiriza", Yellow Kaiser
Near Uwari peninsula, D. R. Congo, W, B, 16 cm
photo: P. Schupke

A97275-4 *Tropheus* sp. Kiriza; Population 1.8
Tropheus "Kiriza", Yellow Kaiser
Near Uwari peninsula, D. R. Congo, W, B, 16 cm
photo: P. Schupke

A97276 *Tropheus* sp. Caramba; Population 1.9
Tropheus "Caramba", Tropheus "Katende"
Cape Muzimo to Cape Caramba, Ubwari peninsula, D. R. Congo, W, B, 16 cm
drawing: P. Schupke

A97276-2 *Tropheus* sp. Caramba; Population 1.9
Tropheus "Caramba", Tropheus "Katende"
Cape Muzimo to Cape Caramba, Ubwari peninsula, D. R. Congo, W, B, 16 cm
photo: P. Schupke

A97276-3 *Tropheus* sp. Caramba; Population 1.9
Tropheus "Caramba", Tropheus "Katende"
Cape Muzimo to Cape Caramba, Ubwari peninsula, D. R. Congo, W, B, 16 cm
photo: P. Schupke

A97276-4 *Tropheus* sp. Caramba; Population 1.9
Tropheus "Caramba", Tropheus "Katende"
Cape Muzimo to Cape Caramba, Ubwari peninsula, D. R. Congo, W, B, 16 cm
photo: P. Schupke

A97276-4 *Tropheus* sp. Caramba; Population 1.9
Tropheus "Caramba", Tropheus "Katende"
Cape Muzimo to Cape Caramba, Ubwari peninsula, D. R. Congo, W, B, 16 cm
photo: P. Schupke

A97276-4 *Tropheus* sp. Caramba; Population 1.9, frightened
Tropheus "Caramba", Tropheus "Katende"
Cape Muzimo to Cape Caramba, Ubwari peninsula, D. R. Congo, W, B, 16 cm
photo: P. Schupke

A97277 *Tropheus brichardi* NELISSEN & THYS, 1975; Population 2.1
Chocolate Moorii, Tropheus "Chocolate", T."Brown", T. "Nyanza Lac I"
Nyanza Lac, Burundi, W, B, 15 cm
drawing: P. Schupke

A97277-4 *Tropheus brichardi* NELISSEN & THYS, 1975; Population 2.1
Chocolate Moorii, Tropheus "Chocolate", T."Brown", T. "Nyanza Lac I"
Nyanza Lac, Burundi, W, B, 15 cm
photo: E. Schraml

A97278 *Tropheus* sp. Gombe; Population 2.2

South of Nyanza Lac, Tanzania, W, 15 cm
drawing: P. Schupke

A97279 *Tropheus* sp. Uvira; Population 2.3

Kavimvira to mouth of Kivovo river, D. R. Congo, W, 15 cm
drawing: P. Schupke

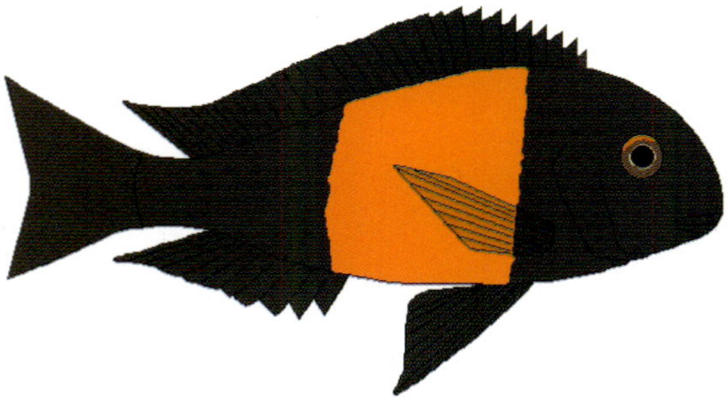

A97280 *Tropheus* sp. Bemba; Population 2.4
Tropheus "Bemba", T. "Orange Band", Orange-Band Moorii
Near Cape Munene, D. R. Congo, W, B, 15 cm
drawing: P. Schupke

A97280-4 *Tropheus* sp. Bemba; Population 2.4, frightened
Tropheus "Bemba", T. "Orange Band", Orange-Band Moorii
Near Cape Munene, D. R. Congo, W, B, 15 cm
photo: P. Schupke

A972804 *Tropheus* sp. Bemba; Population 2.4
Tropheus "Bemba", T. "Orange Band", Orange-Band Moorii
Near Cape Munene, D. R. Congo, W, B, 15 cm
♂ **photo:** P. Schupke

A97280-4 *Tropheus* sp. Bemba; Population 2.4
Tropheus "Bemba", T. "Orange Band", Orange-Band Moorii
Near Cape Munene, D. R. Congo, W, B, 15 cm
♀ **photo:** P. Schupke

A97281 *Tropheus* sp. Makobola I; Population 2.5

From Makobola to the mouth of Kivovo river, D. R. Congo, W, 15 cm
drawing: P. Schupke

A97281-4 *Tropheus* sp. Makobola I; Population 2.5

From Makobola to the mouth of Kivovo river, D. R. Congo, W, 15 cm
photo: W. Staeck

A97282 *Tropheus* sp. Zinzia; Population 3.1

North of Kabogo river and Zinzia creek, D. R. Congo, W, 10-12 cm
drawing: P. Schupke

A97283 *Tropheus* sp. Kabogo; Population 3.2

Between Kabogo river and Muzimo river, D. R. Congo, W, B, 10-12 cm
drawing: P. Schupke

A97283-4 *Tropheus* sp. Kabogo; Population 3.2

Between Kabogo river and Muzimo river, D. R. Congo, W, B, 10-12 cm
♀ **photo:** P. Schupke

A97283-4 *Tropheus* sp. Kabogo; Population 3.2, frightened

Between Kabogo river and Muzimo river, D. R. Congo, W, B, 10-12 cm
♀ **photo:** P. Schupke

A97284 *Tropheus* sp. Kabimba; Population 3.3
Tropheus "Kabimba", Tropheus "Canary Cheek"
Between Muzimo river and Kabimba bay, D. R. Congo, W, B, 10-12 cm
drawing: P. Schupke

A97284-4 *Tropheus* sp. Kabimba; Population 3.3
Tropheus "Kabimba", Tropheus "Canary Cheek"
Between Muzimo river and Kabimba bay, D. R. Congo, W, B, 10-12 cm
♂ **photo:** P. Schupke

A97284-4 *Tropheus* sp. Kabimba; Population 3.3
Tropheus "Kabimba", Tropheus "Canary Cheek"
Between Muzimo river and Kabimba bay, D. R. Congo, W, B, 10-12 cm
♂ **photo:** P. Schupke

A97284-4 *Tropheus* sp. Kabimba; Population 3.3
Tropheus "Kabimba", Tropheus "Canary Cheek"
Between Muzimo river and Kabimba bay, D. R. Congo, W, B, 10-12 cm
♀ **photo:** P. Schupke

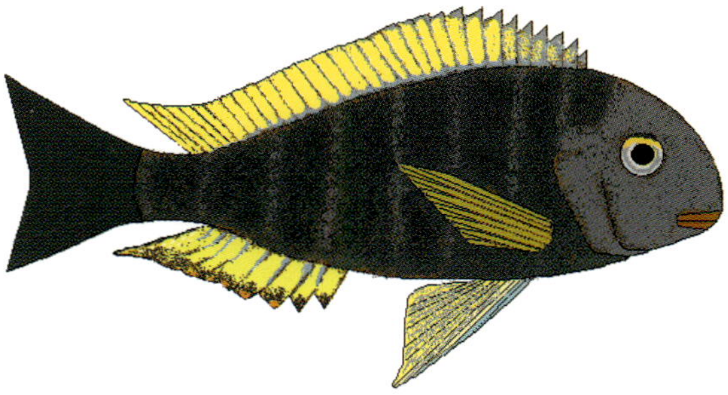

A97285 *Tropheus* sp. Kavala; Population 3.4
Tropheus "Kavala", Tropheus "Kipampa"
Kabimba bay to Cape Bwana Denge, D. R. Congo, W, B, 10-12 cm
drawing: P. Schupke

A97285-4 *Tropheus* sp. Kavala; Population 3.4
Tropheus "Kavala", Tropheus "Kipampa"
Kabimba bay to Cape Bwana Denge, D. R. Congo, W, B, 10-12 cm
♂ **photo:** P. Schupke

A97285-4 *Tropheus* sp. Kavala; Population 3.4
Tropheus "Kavala", Tropheus "Kipampa"
Kabimba bay to Cape Bwana Denge, D. R. Congo, W, B, 10-12 cm
♀ **photo:** P. Schupke

A97286 *Tropheus* sp. Chiwa; Population 3.5
Tropheus "Chiwa", Tropheus "Kipampa"
South of Kabimba bay, D. R. Congo, W, B, 8-9 cm
drawing: P. Schupke

A97286-4 *Tropheus* sp. Chiwa; Population 3.5
Tropheus "Chiwa", Tropheus "Kipampa"
South of Kabimba bay, D. R. Congo, W, B, 8-9 cm
photo: F. Schupke

A97286-4 *Tropheus* sp. Chiwa; Population 3.5
Tropheus "Chiwa", Tropheus "Kipampa"
South of Kabimba bay, D. R. Congo, W, B, 8-9 cm
photo: P. Schupke

A97287 *Tropheus* sp. Kalemie; Population 3.6
Tropheus "Kalemie", T. "Kaniosha", T. "Kioshi"
From Cape Bwana Denge to south from Kalemie, D. R. Congo, W, 10-12 cm
drawing: P. Schupke

A97287-4 *Tropheus* sp. Kalemie; Population 3.6
Tropheus "Kalemie", T. "Kaniosha", T. "Kioshi"
From Cape Bwana Denge to south from Kalemie, D. R. Congo, W, 10-12 cm
♀ **photo:** P. Schupke

A97287-4 *Tropheus* sp. Kalemie; Population 3.6
Tropheus "Kalemie", T. "Kaniosha", T. "Kioshi"
From Cape Bwana Denge to south from Kalemie, D. R. Congo, W, 10-12 cm
♀ **photo:** P. Schupke

A97287-4 *Tropheus* sp. Kalemie; Population 3.6
Tropheus "Kalemie", T. "Kaniosha", T. "Kioshi"
From Cape Bwana Denge to south from Kalemie, D. R. Congo, W, 10-12 cm
♀ **photo:** P. Schupke

A97287-4 *Tropheus* sp. Kalemie; Population 3.6
Tropheus "Kalemie", T. "Kaniosha", T. "Kioshi"
From Cape Bwana Denge to south from Kalemie, D. R. Congo, W, 10-12 cm
photo: W. Staeck

A97287-3 *Tropheus* sp. Kalemie; Population 3.6
Tropheus "Kalemie", T. "Kaniosha", T. "Kioshi"
From Cape Bwana Denge to south from Kalemie, D. R. Congo, W, 10-12 cm
photo: W. Staeck

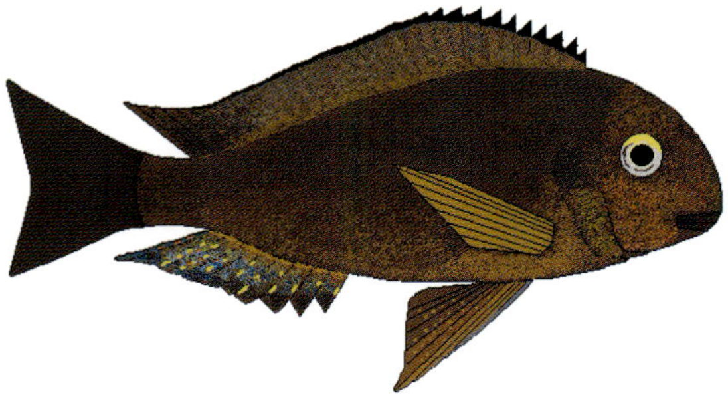

A97288 *Tropheus* sp. Tembwe; Population 3.7

Cape Tembwe, D. R. Congo, W, 10-12 cm

drawing: P. Schupke

A97288-4 *Tropheus* sp. Tembwe; Population 3.7

Cape Tembwe, D. R. Congo, W, 10-12 cm

photo: A. Konings

A97289 *Tropheus* sp. M´toto; Population 3.8
Tropheus "Mtoto", T. "Mikonga", T. "Yellow Nose"
Near M´toto and Pala, D. R. Congo, W, B, 10-12 cm

drawing: P. Schupke

A97289-4 *Tropheus* sp. M´toto; Population 3.8
Tropheus "Mtoto", T. "Mikonga", T. "Yellow Nose"
Near M´toto and Pala, D. R. Congo, W, B, 10-12 cm

photo: A. Konings

A97290 *Tropheus* sp. Karilani; Population 3.9
Tropheus "Karilani"
From Kungwe mountains to Halembe, Tanzania, W, B, 10-12 cm

drawing: P. Schupke

A972904 *Tropheus* sp. Karilani; Population 3.9
Tropheus "Karilani"
From Kungwe mountains to Halembe, Tanzania, W, B, 10-12 cm

photo: P. Schupke

A97291 *Tropheus* sp. Malagarasi; Population 3.10

North of the Malagarasi delta, Tanzania, W, 10-12 cm

drawing: P. Schupke

A97292 *Tropheus* sp. Ujiji (Katonga); Population 3.11
Tropheus "Ujiji", Tropheus "Katonga"
South of Kigoma, Tanzania, W, B, 10-12 cm

drawing: P. Schupke

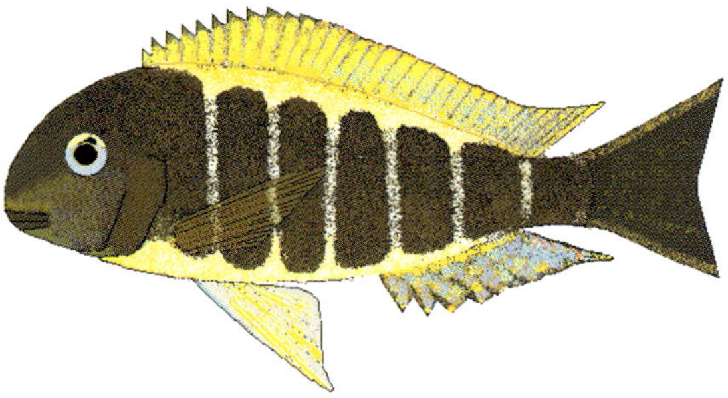

A97292 *Tropheus* sp. Ujiji (Katonga); Population 3.12
Tropheus "Ujiji", Tropheus "Katonga"
SNear Katonga, Tanzania, W, B, 10-12 cm
drawing: P. Schupke

A97292-4 *Tropheus* sp. Ujiji (Katonga); Population 3.11
Tropheus "Ujiji", Tropheus "Katonga"
South of Kigoma, Tanzania, W, B, 10-12 cm
♂ **photo:** P. Schupke

A97292-4 *Tropheus* sp. Ujiji (Katonga); Population 3.11
Tropheus "Ujiji", Tropheus "Katonga"
South of Kigoma, Tanzania, W, B, 10-12 cm
♀ **photo:** P. Schupke

A97292-4 *Tropheus* sp. Ujiji (Katonga); Population 3.11
Tropheus "Ujiji", Tropheus "Katonga"
South of Kigoma, Tanzania, W, B, 10-12 cm
♀ **photo:** P. Schupke

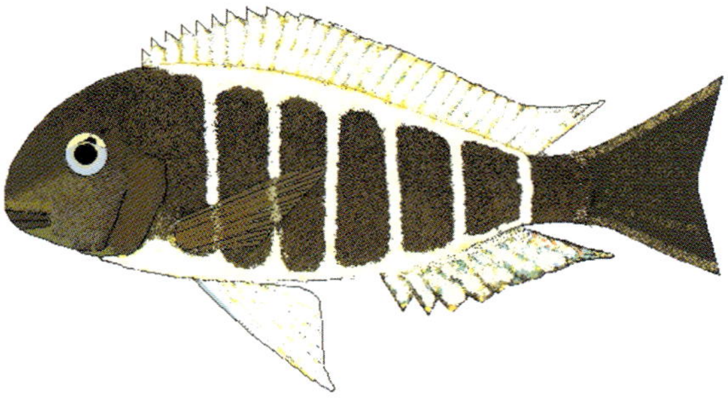

A97293 *Tropheus* sp. Kigoma I (Nyanza Lac II); Population 3.13
Tropheus "Kigoma", Banded Tropheus
South of Kigoma to Nyanza Lac, Tanzania, Burundi, W, B, 10-12 cm
drawing: P. Schupke

A97293 *Tropheus* sp. Kigoma I (Nyanza Lac II); Population 3.14
Tropheus "Kigoma", Banded Tropheus
South of Kigoma to Nyanza Lac, Tanzania, Burundi, W, B, 10-12 cm
drawing: P. Schupke

A97293-4 *Tropheus* sp. Kigoma I (Nyanza Lac II); Population 3.13
Tropheus "Kigoma", Banded Tropheus
South of Kigoma to Nyanza Lac, Tanzania, Burundi, W, B, 10-12 cm
♂ **photo:** P. Schupke

A97293-4 *Tropheus* sp. Kigoma I (Nyanza Lac II); Population 3.13
Tropheus "Kigoma", Banded Tropheus
South of Kigoma to Nyanza Lac, Tanzania, Burundi, W, B, 10-12 cm
♂ **photo:** P. Schupke

A97294 *Tropheus* sp. Bulombora, Population 3.15
Tropheus "Bulombora"
Somewhere between Malagarasi river and Kigoma, Tanzania, W, 10-12 cm
♂ **drawing:** P. Schupke

A97294 *Tropheus* sp. Bulombora, Population 3.15
Tropheus "Bulombora"
Somewhere between Malagarasi river and Kigoma, Tanzania, W, 10-12 cm
♀ **drawing:** P. Schupke

A97294-4 *Tropheus* sp. Bulombora, Population 3.15
Tropheus "Bulombora"
Somewhere between Malagarasi river and Kigoma, Tanzania, W, 10-12 cm
♀ **photo:** P. Schupke

A97294-4 *Tropheus* sp. Bulombora, Population 3.15
Tropheus "Bulombora"
Somewhere between Malagarasi river and Kigoma, Tanzania, W, 10-12 cm
♀ **photo:** P. Schupke

A97294-2 *Tropheus* sp. Bulombora, Population 3.15
Tropheus "Bulombora"
Somewhere between Malagarasi river and Kigoma, Tanzania, W, 10-12 cm
♀ **photo:** P. Schupke

A97294-3 *Tropheus* sp. Bulombora, Population 3.15
Tropheus "Bulombora"
Somewhere between Malagarasi river and Kigoma, Tanzania, W, 10-12 cm
♀ **photo:** P. Schupke

A97294-4 *Tropheus* sp. Bulombora, Population 3.15
Tropheus "Bulombora"
Somewhere between Malagarasi river and Kigoma, Tanzania, W, 10-12 cm
♀ **photo:** P. Schupke

A97295 *Tropheus polli* AXELROD, 1977 Population 4.1
Tropheus polli, Fork-tailed Tropheus
North of Kungwe mountains to near Ikola, Tanzania, W, B, 15-20 cm
drawing: P. Schupke

A97295-4 *Tropheus polli* AXELROD, 1977 Population 4.1
Tropheus polli, Fork-tailed Tropheus
North of Kungwe mountains to near Ikola, Tanzania, W, B, 15-20 cm
photo: P. Schupke

A97295-4 *Tropheus polli* AXELROD, 1977 Population 4.1
Tropheus polli, Fork-tailed Tropheus
North of Kungwe mountains to near Ikola, Tanzania, W, B, 15-20 cm
photo: P. Schupke

A97295-4 *Tropheus polli* AXELROD, 1977 Population 4.1
Tropheus polli, Fork-tailed Tropheus
North of Kungwe mountains to near Ikola, Tanzania, W, B, 15-20 cm
photo: P. Schupke

A97295-5 *Tropheus polli* AXELROD, 1977 Population 4.1
Tropheus polli, Fork-tailed Tropheus
North of Kungwe mountains to near Ikola, Tanzania, W, B, 15-20 cm
photo: P. Schupke

A97296 *Tropheus* sp. Moba I; Population 4.2
Tropheus "Moba"
Moba bay north to Pala, D. R. Congo, W, B, 15-20 cm
drawing: P. Schupke

A97296-3 *Tropheus* sp. Moba I; Population 4.2
Tropheus "Moba"
Moba bay north to Pala, D. R. Congo, W, B, 15-20 cm
photo: P. Schupke

A97296-3 *Tropheus* sp. Moba I; Population 4.2
Tropheus "Moba"
Moba bay north to Pala, D. R. Congo, W, B, 15-20 cm
photo: P. Schupke

A97296-3 *Tropheus* sp. Moba I; Population 4.2
Tropheus "Moba"
Moba bay north to Pala, D. R. Congo, W, B, 15-20 cm
photo: P. Schupke

A97297 *Tropheus* sp. Kyeso; Population 4.3

From Kyeso to south of M´toto, D. R. Congo, W, B, 15-20 cm

drawing: P. Schupke

A97297-4 *Tropheus* sp. Kyeso; Population 4.3

From Kyeso to south of M´toto, D. R. Congo, W, B, 15-20 cm

photo: P. Schupke

A97298 *Tropheus* sp. Kavalla "Kongole"; Population 4.4

Kavalla islands, D. R. Congo, W, B, 15-20 cm

drawing: P. Schupke

A97298-3 *Tropheus* sp. Kavalla "Kongole"; Population 4.4

Kavalla islands, D. R. Congo, W, B, 15-20 cm

photo: P. Schupke

A97298-4 *Tropheus* sp. Kavalla "Kongole"; Population 4.4

Kavalla islands, D. R. Congo, W, B, 15-20 cm

photo: P. Schupke

A97298-4 *Tropheus* sp. Kavalla "Kongole"; Population 4.4

Kavalla islands, D. R. Congo, W, B, 15-20 cm

photo: P. Schupke

A97299 *Tropheus* sp. Kavimvira; Population 5.1
Tropheus "Kavimvira green"
Kavimvira, D. R. Congo, W, B, 10-12 cm

drawing: P. Schupke

A97300 *Tropheus* sp. Makobola II; Population 5.2
Tropheus "Makobola"
Mabokola and further north, D. R. Congo, W, B, 10-12 cm

drawing: P. Schupke

A97301 *Tropheus* sp. Ubwari; Population 5.3
Tropheus "Ubwari green"
Cape Banza to Cape Karamaba, D. R. Congo, W, B, 10-12 cm
drawing: P. Schupke

A97301 *Tropheus* sp. Ubwari; Population 5.4
Tropheus "Ubwari green"
Cape Banza to Cape Karamaba, D. R. Congo, W, B, 10-12 cm
drawing: P. Schupke

A97301-4 *Tropheus* sp. Ubwari; Population 5.3
Tropheus "Ubwari green"
Cape Banza to Cape Karamaba, D. R. Congo, W, B, 10-12 cm
♀ **photo:** P. Schupke

A97301-4 *Tropheus* sp. Ubwari; Population 5.3
Tropheus "Ubwari green"
Cape Banza to Cape Karamaba, D. R. Congo, W, B, 10-12 cm
♀ **photo:** P. Schupke

A97301-4 *Tropheus* sp. Ubwari; Population 5.3
Tropheus "Ubwari green"
Cape Banza to Cape Karamaba, D. R. Congo, W, B, 10-12 cm
♀ **photo:** P. Schupke

A97302 *Tropheus* sp. Kasimia; Population 5.5

Kasimia bay, D. R. Congo, W, B, 10-12 cm
♀ **drawing:** P. Schupke

A97302-4 *Tropheus* sp. Kasimia; Population 5.5

Kasimia bay, D. R. Congo, W, B, 10-12 cm
♂ **photo:** P. Schupke

A97302-4 *Tropheus* sp. Kasimia; Population 5.5

Kasimia bay, D. R. Congo, W, B, 10-12 cm
♀ **photo:** P. Schupke

A97302-4 *Tropheus* sp. Kasimia; Population 5.5

Kasimia bay, D. R. Congo, W, B, 10-12 cm

photo: P. Schupke

A97303 *Tropheus sp.* Yungu; Population 5.6
Tropheus "Yungu"
Near Yungu, D. R. Congo, W, B, 10-12 cm

drawing: P. Schupke

A97303 *Tropheus sp.* Yungu; Population 5.7
Charcoal Moorii
Exact place of origin unknown, D. R. Congo, W, B, 10-12 cm

drawing: P. Schupke

A97303-4 *Tropheus sp.* Yungu; Population 5.6
Tropheus "Yungu"
Near Yungu, D. R. Congo, W, B, 10-12 cm

♂ **photo:** P. Schupke

A97303-4 *Tropheus sp.* Yungu; Population 5.6
Tropheus "Yungu"
Near Yungu, D. R. Congo, W, B, 10-12 cm

♂ **photo:** P. Schupke

A97303-4 *Tropheus sp.* Yungu; Population 5.6
Tropheus "Yungu"
Near Yungu, D. R. Congo, W, B, 10-12 cm

♀ **photo:** P. Schupke

A97303-4 *Tropheus sp.* Yungu; Population 5.6
Tropheus "Yungu"
Near Yungu, D. R. Congo, W, B, 10-12 cm

♀ **photo:** P. Schupke

A97304 *Tropheus* sp. South-east; Population 5.8
Tropheus "Yellow Chest" ("Isonga")
Isonga?, probaly near Halembe, Tanzania, W, B, 10-12 cm

drawing: P. Schupke

A97304-4 *Tropheus* sp. South-east; Population 5.8
Tropheus "Yellow Chest" ("Isonga")
Isonga?, probaly near Halembe, Tanzania, W, B, 10-12 cm
photo: P. Schupke

A97304-3 *Tropheus* sp. South-east; Population 5.8
Tropheus "Yellow Chest" ("Isonga")
Isonga?, probaly near Halembe, Tanzania, W, B, 10-12 cm
photo: P. Schupke

A97304-3 *Tropheus* sp. South-east; Population 5.8
Tropheus "Yellow Chest" ("Isonga")
Isonga?, probaly near Halembe, Tanzania, W, B, 10-12 cm
photo: P. Schupke

A97304-2 *Tropheus* sp. South-east; Population 5.8
Tropheus "Yellow Chest" ("Isonga")
Isonga?, probaly near Halembe, Tanzania, W, B, 10-12 cm
photo: P. Schupke

A97305 *Tropheus* sp. Kabogo; Population 5.9

Cape Kabogo, Tanzania, W, 10-12 cm
drawing: P. Schupke

A97306 *Tropheus* sp. Halembe; Population 5.11

Near Halembe, Tanzania, W, 10-12 cm
drawing: P. Schupke

A97306-4 *Tropheus* sp. Halembe; Population 5.11

Near Halembe, Tanzania, W, 10-12 cm
photo: P. Schupke

A97306-3 *Tropheus* sp. Halembe; Population 5.11

Near Halembe, Tanzania, W, 10-12 cm
photo: P. Schupke

A97306-3 *Tropheus* sp. Halembe; Population 5.11

Near Halembe, Tanzania, W, 10-12 cm

♀ **photo:** P. Schupke

A97905 *Tropheus* sp. Karago; Population 5.10
Tropheus "Malagarasi", Green Moorii, Tropheus "Lugufu"
South of Malagarsi delta to south of Lugufu mouth, Tanzania, W, B, 10-12 cm

drawing: P. Schupke

A97905-4 *Tropheus* sp. Karago; Population 5.10
Tropheus "Malagarasi", Green Moorii, Tropheus "Lugufu"
South of Malagarsi delta to south of Lugufu mouth, Tanzania, W, B, 10-12 cm

♂ **photo:** P. Schupke

A97905-4 *Tropheus* sp. Karago; Population 5.10
Tropheus "Malagarasi", Green Moorii, Tropheus "Lugufu"
South of Malagarsi delta to south of Lugufu mouth, Tanzania, W, B, 10-12 cm

♂ **photo:** P. Schupke

A97905-4 *Tropheus* sp. Karago; Population 5.10
Tropheus "Malagarasi", Green Moorii, Tropheus "Lugufu"
South of Malagarsi delta to south of Lugufu mouth, Tanzania, W, B, 10-12 cm

♀ **photo:** P. Schupke

A97905-4 *Tropheus* sp. Karago; Population 5.10, frightened
Tropheus "Malagarasi", Green Moorii, Tropheus "Lugufu"
South of Malagarsi delta to south of Lugufu mouth, Tanzania, W, B, 10-12 cm

photo: P. Schupke

A97307 *Tropheus* sp. Kigoma II; Population 5.12

North of Kigoma along Ngombe mountains, Tanzania, W, 10-12 cm

drawing: P. Schupke

A97308 *Tropheus* sp. Nyanza III; Population 5.13
Tropheus "Yellow Green"
North of Nyanza Lac, Burundi, W, 10-12 cm

drawing: P. Schupke

A97309 *Tropheus* sp. Rumonge II; Population 5.14

North of Rumonge village, Burundi, W, 10-12 cm

drawing: P. Schupke

A97310 *Tropheus* sp. Kapampa; Population 6.1
Tropheus "Kapampa"
North of Lunangwa to Kapampa bay, D. R. Congo, W, B, 10-12 cm

drawing: P. Schupke

A97309-5 *Tropheus* sp. Kapampa; Population 6.1
Tropheus "Kapampa"
North of Lunangwa to Kapampa bay, D. R. Congo, W, B, 10-12 cm

♂ **photo:** P. Schupke

A97309-4 *Tropheus* sp. Kapampa; Population 6.1
Tropheus "Kapampa"
North of Lunangwa to Kapampa bay, D. R. Congo, W, B, 10-12 cm

♂ **photo:** P. Schupke

A97309-4 *Tropheus* sp. Kapampa; Population 6.1
Tropheus "Kapampa"
North of Lunangwa to Kapampa bay, D. R. Congo, W, B, 10-12 cm

♂ **photo:** P. Schupke

A97309-4 *Tropheus* sp. Kapampa; Population 6.1
Tropheus "Kapampa"
North of Lunangwa to Kapampa bay, D. R. Congo, W, B, 10-12 cm

♀ **photo:** P. Schupke

A97309-4 *Tropheus* sp. Kapampa; Population 6.1
Tropheus "Kapampa"
North of Lunangwa to Kapampa bay, D. R. Congo, W, B, 10-12 cm

♂ ♀ **photo:** P. Schupke

A97309-4 *Tropheus* sp. Kapampa; Population 6.1, frightened
Tropheus "Kapampa"
North of Lunangwa to Kapampa bay, D. R. Congo, W, B, 10-12 cm

♂ ♀ **photo:** P. Schupke

A97311 *Tropheus* sp. Zongwe; Population 6.2
Tropheus "Zongwe"
North of Zongwe village, D. R. Congo, W, B, 10-12 cm
drawing: P. Schupke

A97312 *Tropheus* sp. Kibwesa; Population 6.3

Cape Kibwesa, Tanzania, W, 10-12 cm
drawing: P. Schupke

A97911-5 *Tropheus* sp. Kibwesa; Population 6.3

Cape Kibwesa, Tanzania, W, 10-12 cm
photo: A. Konings

A97313 *Tropheus* sp. Kalya; Population 6.4
Tropheus "Njano"
Kalya, Tanzania, W, B, 10-12 cm
drawing: P. Schupke

A97912-4 *Tropheus* sp. Kalya; Population 6.4
Tropheus "Njano"
Kalya, Tanzania, W, B, 10-12 cm
♂ **photo:** P. Schupke

A97912-4 *Tropheus* sp. Kalya; Population 6.4
Tropheus "Njano"
Kalya, Tanzania, W, B, 10-12 cm
♀ **photo:** P. Schupke

A97912-4 *Tropheus* sp. Kalya; Population 6.4
Tropheus "Njano"
Kalya, Tanzania, W, B, 10-12 cm
♀ **photo:** P. Schupke

A97912-4 *Tropheus* sp. Kalya; Population 6.4
Tropheus "Njano"
Kalya, Tanzania, W, B, 10-12 cm
♀ **photo:** P. Schupke

A97314 *Tropheus* sp. Lunangwa South; Population 7.0

South of Lunangwa, D. R. Congo, W, 10-12 cm

drawing: P. Schupke

A97315 *Tropheus* sp. Lunangwa; Population 7.1
Tropheus "Kiku"
South of Kapampa bay, D. R. Congo, W, B, 10-12 cm

drawing: P. Schupke

A97315-4 *Tropheus* sp. Lunangwa; Population 7.1
Tropheus "Kiku"
South of Kapampa bay, D. R. Congo, W, B, 10-12 cm

photo: P. Schupke

A97315-3 *Tropheus* sp. Lunangwa; Population 7.1
Tropheus "Kiku"
South of Kapampa bay, D. R. Congo, W, B, 10-12 cm

photo: P. Schupke

A97316 *Tropheus* sp. Lupota; Population 7.2
Tropheus "Lupota", Red-Breasted Moorii
South of Lunangwa bay to Lupota bay, D. R. Congo, W, B, 10-12 cm

drawing: P. Schupke

A97316-4 *Tropheus* sp. Lupota; Population 7.2
Tropheus "Lupota", Red-Breasted Moorii
South of Lunangwa bay to Lupota bay, D. R. Congo, W, B, 10-12 cm

photo: P. Schupke

A97316-4 *Tropheus* sp. Lupota; Population 7.2
Tropheus "Lupota", Red-Breasted Moorii
South of Lunangwa bay to Lupota bay, D. R. Congo, W, B, 10-12 cm

photo: P. Schupke

A97317 *Tropheus* sp. Mvua; Population 7.3
Tropheus "Vua"
North of Mvua bay to Lupota bay, D. R. Congo, W, B, 10-12 cm

drawing: P. Schupke

A97317-4 *Tropheus* sp. Mvua; Population 7.3
Tropheus "Vua"
North of Mvua bay to Lupota bay, D. R. Congo, W, B, 10-12 cm
photo: W. Staeck

A97318 *Tropheus* sp. Moliro; Population 7.4
Tropheus "Moliro"
Near Moliro village, D. R. Congo, W, B, 10-12 cm
drawing: P. Schupke

A97318-4 *Tropheus* sp. Moliro; Population 7.4
Tropheus "Moliro"
Near Moliro village, D. R. Congo, W, B, 10-12 cm
photo: P. Schupke

A97318-4 *Tropheus* sp. Moliro; Population 7.4, frightened
Tropheus "Moliro"
Near Moliro village, D. R. Congo, W, B, 10-12 cm
photo: P. Schupke

A97318-2 *Tropheus* sp. Moliro; Population 7.4,
Tropheus "Moliro"
Near Moliro village, D. R. Congo, W, B, 10-12 cm
photo: P. Schupke

A97319 *Tropheus* sp. Chipimbi; Population 7.5
Tropheus "Chipimbi"
Cape Chipimbi, Zambia/D. R. Congo, W, B, 10-12 cm
drawing: P. Schupke

A97319-4 *Tropheus* sp. Chipimbi; Population 7.5
Tropheus "Chipimbi"
Cape Chipimbi, Zambia/D. R. Congo, W, B, 10-12 cm
photo: P. Schupke

A97319-4 *Tropheus* sp. Chipimbi; Population 7.5
Tropheus "Chipimbi"
Cape Chipimbi, Zambia/D. R. Congo, W, B, 10-12 cm
photo: P. Schupke

A97319-4 *Tropheus* sp. Chipimbi; Population 7.5
Tropheus "Chipimbi"
Cape Chipimbi, Zambia/D. R. Congo, W, B, 10-12 cm
photo: P. Schupke

A97320 *Tropheus* sp. Chimba; Population 7.6
Tropheus "Chimba", Red-Black Moorii, Zambian Red
Near Chimba, Cameron bay, Zambia, W, B, 10-12 cm
drawing: P. Schupke

A97320-5 *Tropheus* sp. Chimba; Population 7.6
Tropheus "Chimba", Red-Black Moorii, Zambian Red
Near Chimba, Cameron bay, Zambia, W, B, 10-12 cm
photo: P. Schupke

A97320-5 *Tropheus* sp. Chimba; Population 7.6
Tropheus "Chimba", Red-Black Moorii, Zambian Red
Near Chimba, Cameron bay, Zambia, W, B, 10-12 cm
photo: P. Schupke

A97320-2 *Tropheus* sp. Chimba; Population 7.6
Tropheus "Chimba", Red-Black Moorii, Zambian Red
Near Chimba, Cameron bay, Zambia, W, B, 10-12 cm
photo: P. Schupke

A97320-3 *Tropheus* sp. Chimba; Population 7.6
Tropheus "Chimba", Red-Black Moorii, Zambian Red
Near Chimba, Cameron bay, Zambia, W, B, 10-12 cm
photo: P. Schupke

A97321 *Tropheus* sp. Ndole; Population 7.7
Tropheus "Ndole"
Ndole bay, Zambia, W, B, 10-12 cm
drawing: P. Schupke

A97321-4 *Tropheus* sp. Ndole; Population 7.7
Tropheus "Ndole"
Ndole bay, Zambia, W, B, 10-12 cm
photo: P. Schupke

A97321-4 *Tropheus* sp. Ndole; Population 7.7
Tropheus "Ndole"
Ndole bay, Zambia, W, B, 10-12 cm
photo: P. Schupke

A97322 *Tropheus* sp. Kachese; Population 7.8
Tropheus "Kachese"
Cape Kachese, Zambia, W, B, 10-12 cm
drawing: P. Schupke

A97322-4 *Tropheus* sp. Kachese; Population 7.8
Tropheus "Kachese"
Cape Kachese, Zambia, W, B, 10-12 cm
photo: P. Schupke

A97322-4 *Tropheus* sp. Kachese; Population 7.8
Tropheus "Kachese"
Cape Kachese, Zambia, W, B, 10-12 cm
♂
photo: P. Schupke

A97322-4 *Tropheus* sp. Kachese; Population 7.8
Tropheus "Kachese"
Cape Kachese, Zambia, W, B, 10-12 cm
photo: P. Schupke

A97323 *Tropheus* sp. Sumbu; Population 7.9
Tropheus "Sumbu", Golden Rainbow Moorii
Sumbu bay, Zambia, W, B, 10-12 cm
drawing: P. Schupke

A97323-4 *Tropheus* sp. Sumbu; Population 7.9
Tropheus "Sumbu", Golden Rainbow Moorii
Sumbu bay, Zambia, W, B, 10-12 cm
photo: P. Schupke

A97323-4 *Tropheus* sp. Sumbu; Population 7.9
Tropheus "Sumbu", Golden Rainbow Moorii
Sumbu bay, Zambia, W, B, 10-12 cm
photo: P. Schupke

A97324 *Tropheus* sp. Cape Nundo; Population 7.10
Tropheus "Golden Chisanse"
Cape Nundo, Nkamba bay, Zambia, W, B, 10-12 cm
drawing: P. Schupke

A97324-4 *Tropheus* sp. Cape Nundo; Population 7.10
Tropheus "Golden Chisanse"
Cape Nundo, Nkamba bay, Zambia, W, B, 10-12 cm
photo: P. Schupke

A97324-4 *Tropheus* sp. Cape Nundo; Population 7.10
Tropheus "Golden Chisanse"
Cape Nundo, Nkamba bay, Zambia, W, B, 10-12 cm
photo: P. Schupke

A97324-4 *Tropheus* sp. Cape Nundo; Population 7.10
Tropheus "Golden Chisanse"
Cape Nundo, Nkamba bay, Zambia, W, B, 10-12 cm
photo: P. Schupke

A97325 *Tropheus* sp. Chilanga; Population 7.11
Tropheus "Chilanga Red", Green-Red Moorii, T. "Nkamba"
Nkamba Bay, Zambia, W, B, 10-12 cm
drawing: P. Schupke

A97325-4 *Tropheus* sp. Chilanga; Population 7.11
Tropheus "Chilanga Red", Green-Red Moorii, T. "Nkamba"
Nkamba Bay, Zambia, W, B, 10-12 cm
photo: P. Schupke

A97325-3 *Tropheus* sp. Chilanga; Population 7.11
Tropheus "Chilanga Red", Green-Red Moorii, T. "Nkamba"
Nkamba Bay, Zambia, W, B, 10-12 cm
photo: P. Schupke

A97325-4 *Tropheus* sp. Chilanga; Population 7.11
Tropheus "Chilanga Red", Green-Red Moorii, T. "Nkamba"
Nkamba Bay, Zambia, W, B, 10-12 cm
photo: P. Schupke

A97325-3 *Tropheus* sp. Chilanga; Population 7.11
Tropheus "Chilanga Red", Green-Red Moorii, T. "Nkamba"
Nkamba Bay, Zambia, W, B, 10-12 cm

photo: P. Schupke

A97326 *Tropheus* sp. Ilangi; Population 7.12
Tropheus "Ilangi"
East of Nkamba bay, Zambia, W, B, 10-12 cm

drawing: P. Schupke

A97326-4 *Tropheus* sp. Ilangi; Population 7.12
Tropheus "Ilangi"
East of Nkamba bay, Zambia, W, B, 10-12 cm

photo: P. Schupke

A97326-4 *Tropheus* sp. Ilangi; Population 7.12
Tropheus "Ilangi"
East of Nkamba bay, Zambia, W, B, 10-12 cm

photo: P. Schupke

A97326-4 *Tropheus* sp. Ilangi; Population 7.12
Tropheus "Ilangi"
East of Nkamba bay, Zambia, W, B, 10-12 cm

photo: P. Schupke

A97327 *Tropheus* sp. Linangu; Population 7.13
Tropheus "Linangu"
Southern Kala bay, Zambia, W, B, 10-12 cm

drawing: P. Schupke

A97327-2 *Tropheus* sp. Linangu; Population 7.13
Tropheus "Linangu"
Southern Kala bay, Zambia, W, B, 10-12 cm

photo: P. Schupke

A97327-4 *Tropheus* sp. Linangu; Population 7.13
Tropheus "Linangu"
Southern Kala bay, Zambia, W, B, 10-12 cm

photo: P. Schupke

A97327-4 *Tropheus* sp. Linangu; Population 7.13
Tropheus "Linangu"
Southern Kala bay, Zambia, W, B, 10-12 cm

photo: P. Schupke

A97327-4 *Tropheus* sp. Linangu; Population 7.13
Tropheus "Linangu"
Southern Kala bay, Zambia, W, B, 10-12 cm

photo: P. Schupke

A97327-3 *Tropheus* sp. Linangu; Population 7.13
Tropheus "Linangu"
Southern Kala bay, Zambia, W, B, 10-12 cm

photo: P. Schupke

A97328 *Tropheus* sp. Nangu; Population 7.14
Tropheus "Nangu", Tropheus "Yellow Rainbow"
Inangu peninsula, Zambia, W, B, 10-12 cm

drawing: P. Schupke

A97328-4 *Tropheus* sp. Nangu; Population 7.14
Tropheus "Nangu", Tropheus "Yellow Rainbow"
Inangu peninsula, Zambia, W, B, 10-12 cm

photo: P. Schupke

A97329 *Tropheus* sp. Moba II (Murago); Population 8.1
Tropheus "Murago"
Mola village, D. R. Congo, W, B, 10-12 cm

drawing: P. Schupke

A97329-4 *Tropheus* sp. Moba II (Murago); Population 8.1
Tropheus "Murago"
Mola village, D. R. Congo, W, B, 10-12 cm ♂

photo: P. Schupke

A97329-4 *Tropheus* sp. Moba II (Murago); Population 8.1
Tropheus "Murago"
Mola village, D. R. Congo, W, B, 10-12 cm ♀

photo: P. Schupke

A97329-4 *Tropheus* sp. Moba II (Murago); Population 8.1
Tropheus "Murago"
Mola village, D. R. Congo, W, B, 10-12 cm

photo: P. Schupke

A97330 *Tropheus* sp. Mwerazi North; Population 8.2
Tropheus "Mwerazi"
Near Zongwe village, D. R. Congo, W, B, 10-12 cm

drawing: P. Schupke

A97330-4 *Tropheus* sp. Mwerazi North; Population 8.2
Tropheus "Mwerazi"
Near Zongwe village, D. R. Congo, W, B, 10-12 cm

photo: P. Schupke

A97330-4 *Tropheus* sp. Mwerazi North; Population 8.2
Tropheus "Mwerazi"
Near Zongwe village, D. R. Congo, W, B, 10-12 cm

photo: P. Schupke

A97331 *Tropheus* sp. Mwerazi-Süd; Population 8.3
Tropheus "Mwerazi"
Mwerazi, D. R. Congo, W, B, 10-12 cm

drawing: P. Schupke

A97331 *Tropheus* sp. Mwerazi; Population 8.0
Tropheus "Mwerazi"
Mwerazi, D. R. Congo, W, B, 10-12 cm

drawing: P. Schupke

A97331-4 *Tropheus* sp. Mwerazi-Süd; Population 8.3
Tropheus "Mwerazi"
Mwerazi, D. R. Congo, W, B, 10-12 cm

photo: P. Schupke

A97331-4 *Tropheus* sp. Mwerazi-Süd; Population 8.3
Tropheus "Mwerazi"
Mwerazi, D. R. Congo, W, B, 10-12 cm

photo: P. Schupke

A97332 *Tropheus* sp. Kabeyeye; Population 8.4
Yellow Rainbow, Tropheus "Nangu", Tropheus "Kabeyeye"
Inangu peninsula, Zambia, W, B, 10-12 cm
Drawing: P. Schupke

A97332-4 *Tropheus* sp. Kabeyeye; Population 8.4
Yellow Rainbow, Tropheus "Nangu", Tropheus "Kabeyeye"
Inangu peninsula, Zambia, W, B, 10-12 cm
♂ **photo:** P. Schupke

A97332-4 *Tropheus* sp. Kabeyeye; Population 8.4
Yellow Rainbow, Tropheus "Nangu", Tropheus "Kabeyeye"
Inangu peninsula, Zambia, W, B, 10-12 cm
♀ **photo:** P. Schupke

A97332-4 *Tropheus* sp. Kabeyeye; Population 8.4
Yellow Rainbow, Tropheus "Nangu", Tropheus "Kabeyeye"
Inangu peninsula, Zambia, W, B, 10-12 cm
photo: P. Schupke

A97333 *Tropheus* sp. Lufubu; Population 8.5
Tropheus "Lufubu"
East of Lufubu eiver mouth, Zambia, W, B, 10-12 cm
drawing: P. Schupke

A97333-4 *Tropheus* sp. Lufubu; Population 8.5
Tropheus "Lufubu"
East of Lufubu eiver mouth, Zambia, W, B, 10-12 cm
♀ **photo:** P. Schupke

A97333-4 *Tropheus* sp. Lufubu; Population 8.5
Tropheus "Lufubu"
East of Lufubu eiver mouth, Zambia, W, B, 10-12 cm
photo: P. Schupke

A97333-4 *Tropheus* sp. Lufubu; Population 8.5
Tropheus "Lufubu"
East of Lufubu eiver mouth, Zambia, W, B, 10-12 cm
♂ **photo:** P. Schupke

A97333-3 *Tropheus* sp. Lufubu; Population 8.5
Tropheus "Lufubu"
East of Lufubu eiver mouth, Zambia, W, B, 10-12 cm
photo: P. Schupke

A97333-4 *Tropheus* sp. Lufubu; Population 8.5
Tropheus "Lufubu"
East of Lufubu eiver mouth, Zambia, W, B, 10-12 cm
photo: P. Schupke

A97334 *Tropheus* sp. Chaitika; Population 8.6
Blue Rainbow Moorii, Tropheus "Chaitika"
Cape Chaitika, Zambia, W, B, 10-12 cm
Drawing: P. Schupke

A97334-4 *Tropheus* sp. Chaitika; Population 8.6
Blue Rainbow Moorii, Tropheus "Chaitika"
Cape Chaitika, Zambia, W, B, 10-12 cm
photo: P. Schupke

A97334-4 *Tropheus* sp. Chaitika; Population 8.6
Blue Rainbow Moorii, Tropheus "Chaitika"
Cape Chaitika, Zambia, W, B, 10-12 cm
photo: P. Schupke

A97335 *Tropheus* sp. Kaku; Population 8.7
Tropheus "Kaku"
South of Cape Chaitika, Zambia, W, B, 10-12 cm
Drawing: P. Schupke

A97336 *Tropheus* sp. Katoto; Population 8.8
Tropheus "Katoto"
East of Mbete, Zambia, W, B, 10-12 cm
Drawing: P. Schupke

A97336-4 *Tropheus* sp. Katoto; Population 8.8
Tropheus "Katoto"
East of Mbete, Zambia, W, B, 10-12 cm
photo: L. DeMason

A97336-4 *Tropheus* sp. Katoto; Population 8.8
Tropheus "Katoto"
East of Mbete, Zambia, W, B, 10-12 cm

photo: A. Konings

A97337 *Tropheus moorii* BOULENGER, 1898; Population 8.9, Mbete
Tropheus "Mbita", Lemon Spot Moorii, T. "Yellow Spot"
Near Mpulungu, Zambia, W, B, 10-12 cm

drawing: P. Schupke

A97337-4 *Tropheus moorii* BOULENGER, 1898; Population 8.9, Mbete
Tropheus "Mbita", Lemon Spot Moorii, T. "Yellow Spot"
Near Mpulungu, Zambia, W, B, 10-12 cm

photo: P. Schupke

A97337-4 *Tropheus moorii* BOULENGER, 1898; Population 8.9, Mbete
Tropheus "Mbita", Lemon Spot Moorii, T. "Yellow Spot"
Near Mpulungu, Zambia, W, B, 10-12 cm

drawing: P. Schupke

A97338 *Tropheus* sp. Mutondwe; Population 8.10
Tropheus "Mutondwe"
Mutondwe island, Zambia, W, B, 10-12 cm

drawing: P. Schupke

A97338-4 *Tropheus* sp. Mutondwe; Population 8.10
Tropheus "Mutondwe"
Mutondwe island, Zambia, W, B, 10-12 cm

photo: A. Konings

A97339 *Tropheus sp.* Kasakalawe; Population 8.11
Tropheus "Orange Spot", Tropheus "Sonsitt"
Near Kasakalawe, Z,ania, W, B, 10-12 cm

drawing: P. Schupke

A97339-4 *Tropheus sp.* Kasakalawe; Population 8.11
Tropheus "Orange Spot", Tropheus "Sonsitt"
Near Kasakalawe, Z,ania, W, B, 10-12 cm

photo: P. Schupke

A97339-4 *Tropheus sp.* Kasakalawe; Population 8.11
Tropheus "Orange Spot", Tropheus "Sonsitt"
Near Kasakalawe, Z,ania, W, B, 10-12 cm
photo: P. Schupke

A97339-4 *Tropheus sp.* Kasakalawe; Population 8.11
Tropheus "Orange Spot", Tropheus "Sonsitt"
Near Kasakalawe, Z,ania, W, B, 10-12 cm
photo: P. Schupke

A97340 *Tropheus* sp. Kalambo; Population 8.12
Tropheus "Kalambo", Lemon Spot Tropheus
South of Kalambo river to Chituta bay, Zambia, W, B, 10-12 cm
drawing: P. Schupke

A97340-4 *Tropheus* sp. Kalambo; Population 8.12
Tropheus "Kalambo", Lemon Spot Tropheus
South of Kalambo river to Chituta bay, Zambia, W, B, 10-12 cm
♂ **photo:** P. Schupke

A97340-4 *Tropheus* sp. Kalambo; Population 8.12
Tropheus "Kalambo", Lemon Spot Tropheus
South of Kalambo river to Chituta bay, Zambia, W, B, 10-12 cm
♀ **photo:** P. Schupke

A97340-4 *Tropheus* sp. Kalambo; Population 8.12
Tropheus "Kalambo", Lemon Spot Tropheus
South of Kalambo river to Chituta bay, Zambia, W, B, 10-12 cm
♂ **photo:** P. Schupke

A97342 *Tropheus* sp. Kasanga; Population 8.13
Tropheus "Kasanga", Red Rainbow Moorii
North of Kalambo river to Kasanga, Tanzania, W, B, 10-12 cm
drawing: P. Schupke

A97342-4 *Tropheus* sp. Kasanga; Population 8.13
Tropheus "Kasanga", Red Rainbow Moorii
North of Kalambo river to Kasanga, Tanzania, W, B, 10-12 cm
♀ **photo:** P. Schupke

A97342-4 *Tropheus* sp. Kasanga; Population 8.13
Tropheus "Kasanga", Red Rainbow Moorii
North of Kalambo river to Kasanga, Tanzania, W, B, 10-12 cm

photo: P. Schupke

A97342-4 *Tropheus* sp. Kasanga; Population 8.13. miscoloured
Tropheus "Kasanga", Red Rainbow Moorii
North of Kalambo river to Kasanga, Tanzania, W, B, 10-12 cm

photo: P. Schupke

A97342-4 *Tropheus* sp. Kasanga; Population 8.13
Tropheus "Kasanga", Red Rainbow Moorii
North of Kalambo river to Kasanga, Tanzania, W, B, 10-12 cm

photo: P. Schupke

A97342-4 *Tropheus* sp. Kasanga; Population 8.13
Tropheus "Kasanga", Red Rainbow Moorii
North of Kalambo river to Kasanga, Tanzania, W, B, 10-12 cm

photo: P. Schupke

A97343 *Tropheus* sp. Muzi; Population 8.14
Tropheus "Muzi", Red Rainbow Blue Blaze, Blue Kasanga Moorii
Near Muzi village, Tanzania, W, B, 10-12 cm

drawing: P. Schupke

A97343-4 *Tropheus* sp. Muzi; Population 8.14
Tropheus "Muzi", Red Rainbow Blue Blaze, Blue Kasanga Moorii
Near Muzi village, Tanzania, W, B, 10-12 cm

photo: P. Schupke

A97343-4 *Tropheus* sp. Muzi; Population 8.14
Tropheus "Muzi", Red Rainbow Blue Blaze, Blue Kasanga Moorii
Near Muzi village, Tanzania, W, B, 10-12 cm

photo: P. Schupke

A97343-4 *Tropheus* sp. Muzi; Population 8.14
Tropheus "Muzi", Red Rainbow Blue Blaze, Blue Kasanga Moorii
Near Muzi village, Tanzania, W, B, 10-12 cm

photo: P. Schupke

A97344 *Tropheus* sp. Molwe; Population 8.15
Tropheus "Cherry Rainbow"
North of Molwe village, Tanzania, W, 10-12 cm

drawing: P. Schupke

A97344-4 *Tropheus* sp. Molwe; Population 8.15
Tropheus "Cherry Rainbow"
North of Molwe village, Tanzania, W, 10-12 cm

photo: P. Schupke

A97345 *Tropheus* sp. Mamalesa (Malasa) Population 8.16
Tropheus "Malasa"
Mamalesa islands, Tanzania, W, B, 10-12 cm

drawing: P. Schupke

A97345-4 *Tropheus* sp. Mamalesa (Malasa) Population 8.16
Tropheus "Malasa"
Mamalesa islands, Tanzania, W, B, 10-12 cm

♂ **photo:** P. Schupke

A97345-4 *Tropheus* sp. Mamalesa (Malasa) Population 8.16
Tropheus "Malasa"
Mamalesa islands, Tanzania, W, B, 10-12 cm

♀ **photo:** P. Schupke

A97346 *Tropheus* sp. Kala; Population 8.17
Tropheus "Kala Yellow"
Kala island, Tanzania, W, B, 10-12 cm

drawing: P. Schupke

A97346-4 *Tropheus* sp. Kala; Population 8.17
Tropheus "Kala Yellow"
Kala island, Tanzania, W, B, 10-12 cm

photo: P. Schupke

A97346-5 *Tropheus* sp. Kala; Population 8.17
Tropheus "Kala Yellow"
Kala island, Tanzania, W, B, 10-12 cm

photo: P. Schupke

A97346-5 *Tropheus* sp. Kala; Population 8.17
Tropheus "Kala Yellow"
Kala island, Tanzania, W, B, 10-12 cm

photo: P. Schupke

A97347 *Tropheus* sp. Wampembe; Population 8.18
Tropheus "Wampembe"
Near Wampempe village, Tanzania, W, B, 10-12 cm

drawing: P. Schupke

A97347-4 *Tropheus* sp. Wampembe; Population 8.18
Tropheus "Wampembe"
Near Wampempe village, Tanzania, W, B, 10-12 cm

photo: P. Schupke

A97347-4 *Tropheus* sp. Wampembe; Population 8.18
Tropheus "Wampembe"
Near Wampempe village, Tanzania, W, B, 10-12 cm

photo: P. Schupke

A97347-4 *Tropheus* sp. Wampembe; Population 8.18
Tropheus "Wampembe"
Near Wampempe village, Tanzania, W, B, 10-12 cm

photo: P. Schupke

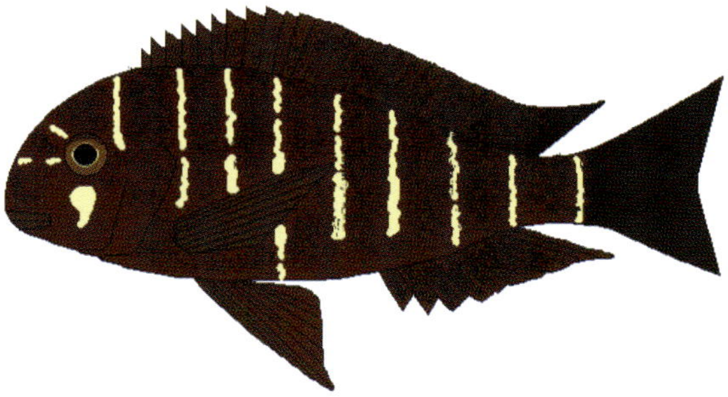

A97348 *Tropheus* sp. Hinde B; Population 8.19

Near Hinde B village, Tanzania, W, 10-12 cm

drawing: P. Schupke

A97348-4 *Tropheus* sp. Hinde B; Population 8.19

Near Hinde B village, Tanzania, W, 10-12 cm

photo: A. Konings

A97349 *Tropheus* sp. Namansi I; Population 8.20
Tropheus "Namansi", Red Namansi
Near Namansi village, Tanzania, W, B, 10-12 cm

drawing: P. Schupke

A97349-4 *Tropheus* sp. Namansi I; Population 8.20
Tropheus "Namansi", Red Namansi
Near Namansi village, Tanzania, W, B, 10-12 cm
photo: P. Schupke

A97349-4 *Tropheus* sp. Namansi I; Population 8.20
Tropheus "Namansi", Red Namansi
Near Namansi village, Tanzania, W, B, 10-12 cm
photo: P. Schupke

A97349-4 *Tropheus* sp. Namansi I; Population 8.20
Tropheus "Namansi", Red Namansi
Near Namansi village, Tanzania, W, B, 10-12 cm
photo: P. Schupke

A97349-4 *Tropheus* sp. Namansi I; Population 8.20
Tropheus "Namansi", Red Namansi
Near Namansi village, Tanzania, W, B, 10-12 cm
photo: P. Schupke

A97350 *Tropheus* sp. Mtosi I; Population 8.21

Mtosi, Tanzania, W, 10-12 cm
drawing: P. Schupke

A97350-4 *Tropheus* sp. Mtosi I; Population 8.21

Mtosi, Tanzania, W, 10-12 cm
photo: A. Konings

A97351 *Tropheus* sp. Kisambali; Population 8.22

Kisambali, Tanzania, W, 10-12 cm
drawing: P. Schupke

A97351-4 *Tropheus* sp. Kisambali; Population 8.22

Kisambali, Tanzania, W, 10-12 cm
photo: A. Konings

A97352 *Tropheus* sp. Namansi II; Population 9.0
Tropheus "Fiery Fry"
Near Namansi village, Tanzania, W, B, 10-12 cm
drawing: P. Schupke

A97352-4 *Tropheus* sp. Namansi II; Population 9.0
Tropheus "Fiery Fry"
Near Namansi village, Tanzania, W, B, 10-12 cm
photo: P. Schupke

A97352-4 *Tropheus* sp. Namansi II; Population 9.0
Tropheus "Fiery Fry"
Near Namansi village, Tanzania, W, B, 10-12 cm
photo: P. Schupke

A97352-2 *Tropheus* sp. Namansi II; Population 9.0
Tropheus "Fiery Fry"
Near Namansi village, Tanzania, W, B, 10-12 cm
photo: P. Schupke

A97352-2 *Tropheus* sp. Namansi II; Population 9.0
Tropheus "Fiery Fry"
Near Namansi village, Tanzania, W, B, 10-12 cm
photo: P. Schupke

A97353 *Tropheus* sp. Mtosi I; Population 9.1
Tropheus "Mtosi"
Near Mtosi, Tanzania, W, B, 10-12 cm
drawing: P. Schupke

A97353 *Tropheus* sp. Mtosi I; Population 9.1
Tropheus "Mtosi"
Near Mtosi, Tanzania, W, B, 10-12 cm
drawing: P. Schupke

A97353-4 sp. Mtosi I; Population 9.1
Tropheus "Mtosi"
Near Mtosi, Tanzania, W, B, 10-12 cm
photo: A. Konings

A97354 *Tropheus* sp. Ulwile; Population 9.2
Tropheus "Ulwile"
Ulwile island, Tanzania, W, B, 10-12 cm
♂ **drawing:** P. Schupke

A97354 *Tropheus* sp. Ulwile; Population 9.2
Tropheus "Ulwile"
Ulwile island, Tanzania, W, B, 10-12 cm
♀ **drawing:** P. Schupke

A97354-4 *Tropheus* sp. Ulwile; Population 9.2
Tropheus "Ulwile"
Ulwile island, Tanzania, W, B, 10-12 cm
♀ **photo:** P. Schupke

A97354-4 *Tropheus* sp. Ulwile; Population 9.2
Tropheus "Ulwile"
Ulwile island, Tanzania, W, B, 10-12 cm
♀ **photo:** P. Schupke

A97355 *Tropheus* sp. Kipili; Population 9.3
Tropheus "Kipili", Wasp Moorii
Near Kipili, Tanzania, W, B, 10-12 cm
♂ **drawing:** P. Schupke

A97355 *Tropheus* sp. Kipili; Population 9.3
Tropheus "Kipili", Wasp Moorii
Near Kipili, Tanzania, W, B, 10-12 cm
♀ **drawing:** P. Schupke

A97355-4 *Tropheus* sp. Kipili; Population 9.3
Tropheus "Kipili", Wasp Moorii
Near Kipili, Tanzania, W, B, 10-12 cm
♂ **photo:** P. Schupke

A97355-4 *Tropheus* sp. Kipili; Population 9.3
Tropheus "Kipili", Wasp Moorii
Near Kipili, Tanzania, W, B, 10-12 cm
♂ **photo:** P. Schupke

A97355-4 *Tropheus* sp. Kipili; Population 9.3
Tropheus "Kipili", Wasp Moorii
Near Kipili, Tanzania, W, B, 10-12 cm
photo: P. Schupke

A97355-2 *Tropheus* sp. Kipili; Population 9.3
Tropheus "Kipili", Wasp Moorii
Near Kipili, Tanzania, W, B, 10-12 cm
photo: P. Schupke

A97356 *Tropheus* sp. Kampemba; Population 9.4
Tropheus "Kampemba"
Cape Kampemba, Tanzania, W, B, 10-12 cm
drawing: P. Schupke

A97356 *Tropheus* sp. Kampemba; Population 9.4
Tropheus "Kampemba"
Cape Kampemba, Tanzania, W, B, 10-12 cm
drawing: P. Schupke

A97356-4 *Tropheus* sp. Kampemba; Population 9.4
Tropheus "Kampemba"
Cape Kampemba, Tanzania, W, B, 10-12 cm
photo: P. Schupke

A97356-4 *Tropheus* sp. Kampemba; Population 9.4
Tropheus "Kampemba"
Cape Kampemba, Tanzania, W, B, 10-12 cm
photo: P. Schupke

A97357 *Tropheus* sp. Korongwe; Population 9.5
Tropheus "Mpimbwe", Tropheus "Dahabu", T. "Msalaba"
South of Cape Mpimbwe, Tanzania, W, B, 10-12 cm
drawing: P. Schupke

A97357 *Tropheus* sp. Korongwe; Population 9.5
Tropheus "Mpimbwe", Tropheus "Dahabu", T. "Msalaba"
South of Cape Mpimbwe, Tanzania, W, B, 10-12 cm
drawing: P. Schupke

A97357-4 *Tropheus* sp. Korongwe; Population 9.5
Tropheus "Mpimbwe", Tropheus "Dahabu", T. "Msalaba"
South of Cape Mpimbwe, Tanzania, W, B, 10-12 cm
photo: P. Schupke

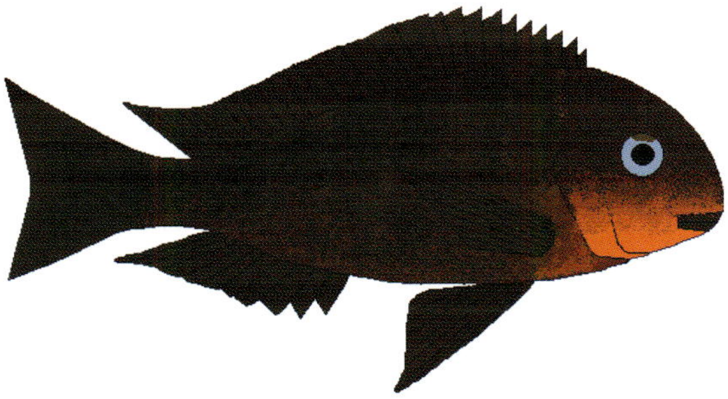

A97358 *Tropheus* sp. Mpimbwe; Population 9.6
Tropheus "Mpimbwe", Tropheus "Dahabu", T. "Msalaba"
Cape Mpimbwe, Tanzania, W, B, 10-12 cm
drawing: P. Schupke

A97358 *Tropheus* sp. Mpimbwe; Population 9.6
Tropheus "Mpimbwe", Tropheus "Dahabu", T. "Msalaba"
Cape Mpimbwe, Tanzania, W, B, 10-12 cm
drawing: P. Schupke

A97358-4 *Tropheus* sp. Mpimbwe; Population 9.6
Tropheus "Mpimbwe", Tropheus "Dahabu", T. "Msalaba"
Cape Mpimbwe, Tanzania, W, B, 10-12 cm
photo: P. Schupke

A97358-4 *Tropheus* sp. Mpimbwe; Population 9.6
Tropheus "Mpimbwe", Tropheus "Dahabu", T. "Msalaba"
Cape Mpimbwe, Tanzania, W, B, 10-12 cm
photo: P. Schupke

A97358-4 *Tropheus* sp. Mpimbwe; Population 9.6
Tropheus "Mpimbwe", Tropheus "Dahabu", T. "Msalaba"
Cape Mpimbwe, Tanzania, W, B, 10-12 cm
photo: P. Schupke

A97358-4 *Tropheus* sp. Mpimbwe; Population 9.6
Tropheus "Mpimbwe", Tropheus "Dahabu", T. "Msalaba"
Cape Mpimbwe, Tanzania, W, B, 10-12 cm
photo: P. Schupke

A97358-3 *Tropheus* sp. Mpimbwe; Population 9.6
Tropheus "Mpimbwe", Tropheus "Dahabu", T. "Msalaba"
Cape Mpimbwe, Tanzania, W, B, 10-12 cm
photo: P. Schupke

A97359 *Tropheus* sp. Kabwe; Population 9.7
Tropheus "Kashekezi", Tropheus "Golden Kushangaza"
Near Kabwe, Tanzania, W, B, 10-12 cm

drawing: P. Schupke

A97360 *Tropheus* sp. Kabwe; Population 9.7, Mutant
Tropheus "Kashekezi", Tropheus "Golden Kushangaza"
Near Kabwe, Tanzania, W, B, 10-12 cm

drawing: P. Schupke

A97359-4 *Tropheus* sp. Kabwe; Population 9.7
Tropheus "Kashekezi", Tropheus "Golden Kushangaza"
Near Kabwe, Tanzania, W, B, 10-12 cm

photo: P. Schupke

A97359-4 *Tropheus* sp. Kabwe; Population 9.7
Tropheus "Kashekezi", Tropheus "Golden Kushangaza"
Near Kabwe, Tanzania, W, B, 10-12 cm

photo: P. Schupke

A97359-3 *Tropheus* sp. Kabwe; Population 9.7
Tropheus "Kashekezi", Tropheus "Golden Kushangaza"
Near Kabwe, Tanzania, W, B, 10-12 cm

photo: P. Schupke

A97360 *Tropheus* sp. Kabwe; Population 9.7, Mutant
Tropheus "Kashekezi", Tropheus "Golden Kushangaza"
Near Kabwe, Tanzania, W, B, 10-12 cm

photo: P. Schupke

A97361 *Tropheus* sp. Kalila; Population 9.8
Tropheus "Kalila Red Eye", Tropheus "Mkombe", T. "Karema"
South of Mkombe to Karema, Tanzania, W, B, 10-12 cm

♂ **drawing:** P. Schupke

A97361 *Tropheus* sp. Kalila; Population 9.8
Tropheus "Kalila Red Eye", Tropheus "Mkombe", T. "Karema"
South of Mkombe to Karema, Tanzania, W, B, 10-12 cm

♀ **drawing:** P. Schupke

A97361-4 *Tropheus* sp. Kalila; Population 9.8
Tropheus "Kalila Red Eye", Tropheus "Mkombe", T. "Karema"
South of Mkombe to Karema, Tanzania, W, B, 10-12 cm
photo: A. Konings

A97361-4 *Tropheus* sp. Kalila; Population 9.8
Tropheus "Kalila Red Eye", Tropheus "Mkombe", T. "Karema"
South of Mkombe to Karema, Tanzania, W, B, 10-12 cm
photo: L. DeMason

A97362 *Tropheus duboisi* MARLIER, 1959; Population 10.1
T. duboisi "White Band", White Band Duboisi, Bemba Duboisi
Cape Munene, D. R. Congo, W, B, 12-15 cm
drawing: P. Schupke

A97362-5 *Tropheus duboisi* MARLIER, 1959; Population 10.1
T. duboisi "White Band", White Band Duboisi, Bemba Duboisi
Cape Munene, D. R. Congo, W, B, 12-15 cm
photo: Archiv A.C.S.

A97362-4 *Tropheus duboisi* MARLIER, 1959; Population 10.1
T. duboisi "White Band", White Band Duboisi, Bemba Duboisi
Cape Munene, D. R. Congo, W, B, 12-15 cm
photo: E. Schraml

A97362-2 *Tropheus duboisi* MARLIER, 1959; Population 10.1
T. duboisi "White Band", White Band Duboisi, Bemba Duboisi
Cape Munene, D. R. Congo, W, B, 12-15 cm
photo: F. Teigler/Archiv A.C.S.

A97363 *Tropheus* cf. *duboisi;* Population 10.2
T. duboisi "Yellow Band", Yellow Band Duboisi, Kigoma Duboisi
Nyanza Lac to the edelta of Malagarasi river, Burundi/Tanzania, W, B, 12-15 cm
drawing: P. Schupke

A97363-1 *Tropheus* cf. *duboisi;* Population 10.2
T. duboisi "Yellow Band", Yellow Band Duboisi, Kigoma Duboisi
Nyanza Lac to the edelta of Malagarasi river, Burundi/Tanzania, W, B, 12-15 cm
photo: P. Schupke

A97363-5 *Tropheus* cf. *duboisi;* Population 10.2
T. duboisi "Yellow Band", Yellow Band Duboisi, Kigoma Duboisi
Nyanza Lac to the edelta of Malagarasi river, Burundi/Tanzania, W, B, 12-15 cm
♂ **photo:** P. Schupke

A97363-5 *Tropheus* cf. *duboisi;* Population 10.2
T. duboisi "Yellow Band", Yellow Band Duboisi, Kigoma Duboisi
Nyanza Lac to the edelta of Malagarasi river, Burundi/Tanzania, W, B, 12-15 cm
♀ **photo:** P. Schupke

A97364 *Tropheus* cf. *duboisi;* Population 10.3
T. duboisi "Broad Band", Broad Band Duboisi, T. duboisi "Maswa"
South of Lugufu river to Cape Kabogo, Tanzania, W, B, 12-15 cm
drawing: P. Schupke

A97364-4 *Tropheus* cf. *duboisi;* Population 10.3
T. duboisi "Broad Band", Broad Band Duboisi, T. duboisi "Maswa"
South of Lugufu river to Cape Kabogo, Tanzania, W, B, 12-15 cm
♂ **photo:** P. Schupke

A97364-3 *Tropheus* cf. *duboisi;* Population 10.3
T. duboisi "Broad Band", Broad Band Duboisi, T. duboisi "Maswa"
South of Lugufu river to Cape Kabogo, Tanzania, W, B, 12-15 cm
♂ **photo:** P. Schupke

A97364-5 *Tropheus* cf. *duboisi;* Population 10.3
T. duboisi "Broad Band", Broad Band Duboisi, T. duboisi "Maswa"
South of Lugufu river to Cape Kabogo, Tanzania, W, B, 12-15 cm
♂ **photo:** P. Schupke

A97365 *Tropheus* cf. *duboisi;* Population 10.4
Tropheus duboisi "Karilani", Karilani Duboisi
Magambo village and Karilani island, Tanzania, W, B, 12-15 cm
drawing: P Schupke

A97365-4 *Tropheus* cf. *duboisi;* Population 10.4
Tropheus duboisi "Karilani", Karilani Duboisi
Magambo village and Karilani island, Tanzania, W, B, 12-15 cm
photo: A. Konings

A97366-3 *Tropheus* cf. *duboisi*
Black Duboisi
Origin unknown, B, 12-15 cm
photo: Richter/Archiv A.C.S.

A97367 *Tropheus* sp. Ikola; Population 11.1
Tropheus sp. "Ikola", Kaiser Moorii
North of Ikola village, Tanzania, W, B, 10-12 cm
drawing: P. Schupke

A97367-4 *Tropheus* sp. Ikola; Population 11.1
Tropheus sp. "Ikola", Kaiser Moorii
North of Ikola village, Tanzania, W, B, 10-12 cm
photo: P. Schupke

A97367-4 *Tropheus* sp. Ikola; Population 11.1
Tropheus sp. "Ikola", Kaiser Moorii
North of Ikola village, Tanzania, W, B, 10-12 cm
photo: P. Schupke

A97367-4 *Tropheus* sp. Ikola; Population 11.1
Tropheus sp. "Ikola", Kaiser Moorii
North of Ikola village, Tanzania, W, B, 10-12 cm
photo: P. Schupke

A97367-4 *Tropheus* sp. Ikola; Population 11.1
Tropheus sp. "Ikola", Kaiser Moorii
North of Ikola village, Tanzania, W, B, 10-12 cm
photo: P. Schupke

A97367-4 *Tropheus* sp. Ikola; Population 11.1
Tropheus sp. "Ikola", Kaiser Moorii
North of Ikola village, Tanzania, W, B, 10-12 cm
photo: P. Schupke

A97368-4 Hybrid of Population 11.1 and Population 12.3
photo: P. Schupke

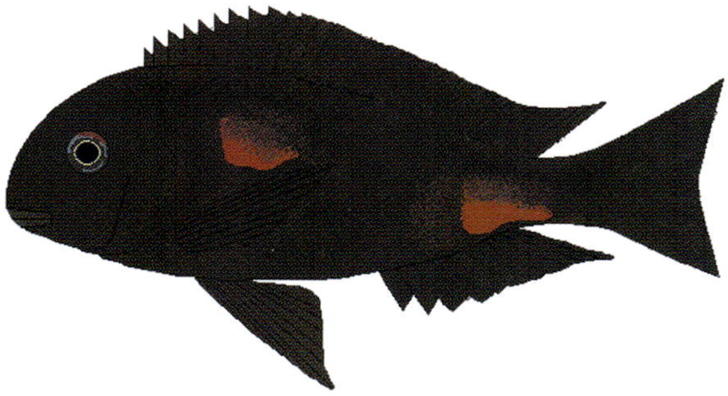

A97369 *Tropheus* sp. Kibwesa Nord; Population 12.0
Tropheus "Kirschfleck", T. "Cherry Spot", Kirschfleck Moorii
Near Cape Kibwesa, Tanzania, W, B, 10-12 cm
drawing: P. Schupke

A97369-4 *Tropheus* sp. Kibwesa Nord; Population 12.0
Tropheus "Kirschfleck", T. "Cherry Spot", Kirschfleck Moorii
Near Cape Kibwesa, Tanzania, W, B, 10-12 cm
photo: P. Schupke

A97369-4 *Tropheus* sp. Kibwesa Nord; Population 12.0
Tropheus "Kirschfleck", T. "Cherry Spot", Kirschfleck Moorii
Near Cape Kibwesa, Tanzania, W, B, 10-12 cm
photo: P. Schupke

A97370 *Tropheus sp. Mabilibili; Population 12.1*
Tropheus "Kirschfleck", T. "Cherry Spot", Kirschfleck Moorii
Near Cape Kibwesa, Tanzania, W, B, 10-12 cm
drawing: P. Schupke

A97370-4 *Tropheus sp. Mabilibili; Population 12.1*
Tropheus "Kirschfleck", T. "Cherry Spot", Kirschfleck Moorii
Near Cape Kibwesa, Tanzania, W, B, 10-12 cm
photo: P. Schupke

A97370-4 *Tropheus sp. Mabilibili; Population 12.1*
Tropheus "Kirschfleck", T. "Cherry Spot", Kirschfleck Moorii
Near Cape Kibwesa, Tanzania, W, B, 10-12 cm
photo: P. Schupke

A97370-4 *Tropheus sp. Mabilibili; Population 12.1*
Tropheus "Kirschfleck", T. "Cherry Spot", Kirschfleck Moorii
Near Cape Kibwesa, Tanzania, W, B, 10-12 cm
photo: P. Schupke

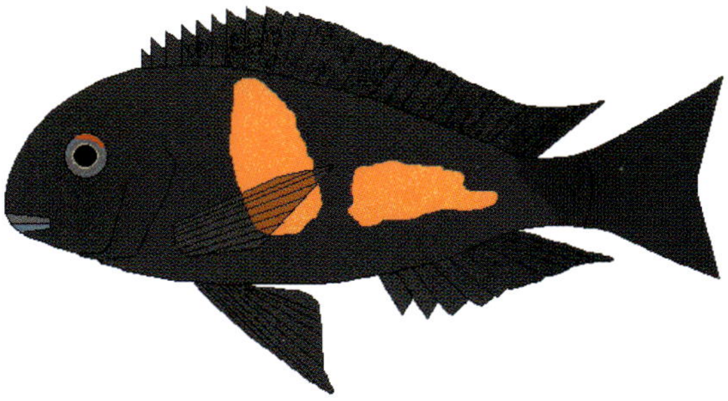

A97371 *Tropheus* sp. Siyeswe; Population 12.2
Tropheus "Kirschfleck", T. "Cherry Spot", Kirschfleck Moorii
Near Siyeswe village, Tanzania, W, B, 10-12 cm
drawing: P. Schupke

A97371 *Tropheus* sp. Siyeswe; Population 12.2
Tropheus "Kirschfleck", T. "Cherry Spot", Kirschfleck Moorii
Near Siyeswe village, Tanzania, W, B, 10-12 cm
photo: P. Schupke

A97371 *Tropheus* sp. Siyeswe; Population 12.2
Tropheus "Kirschfleck", T. "Cherry Spot", Kirschfleck Moorii
Near Siyeswe village, Tanzania, W, B, 10-12 cm
photo: P. Schupke

A97372 *Tropheus* sp. Bulu Point; Population 12.3
Tropheus "Kirschfleck", T. "Cherry Spot", Kirschfleck Moorii
Bulu Pnt, Tanzania, W, B, 10-12 cm
drawing: P. Schupke

A97372 *Tropheus* sp. Bulu Point; Population 12.3
Tropheus "Kirschfleck", T. "Cherry Spot", Kirschfleck Moorii
Bulu Pnt, Tanzania, W, B, 10-12 cm
drawing: P. Schupke

A97372 *Tropheus* sp. Bulu Point; Population 12.3
Tropheus "Kirschfleck", T. "Cherry Spot", Kirschfleck Moorii
Bulu Pnt, Tanzania, W, B, 10-12 cm
drawing: P. Schupke

A97372-4 *Tropheus* sp. Bulu Point; Population 12.3
Tropheus "Kirschfleck", T. "Cherry Spot", Kirschfleck Moorii
Bulu Pnt, Tanzania, W, B, 10-12 cm
photo: P. Schupke

A97372-4 *Tropheus* sp. Bulu Point; Population 12.3
Tropheus "Kirschfleck", T. "Cherry Spot", Kirschfleck Moorii
Bulu Pnt, Tanzania, W, B, 10-12 cm
photo: P. Schupke

A97372-4 *Tropheus* sp. Bulu Point; Population 12.3
Tropheus "Kirschfleck", T. "Cherry Spot", Kirschfleck Moorii
Bulu Pnt, Tanzania, W, B, 10-12 cm
photo: P. Schupke

A97372-4 *Tropheus* sp. Bulu Point; Population 12.3
Tropheus "Kirschfleck", T. "Cherry Spot", Kirschfleck Moorii
Bulu Pnt, Tanzania, W, B, 10-12 cm

photo: P. Schupke

A97372-4 *Tropheus* sp. Bulu Point; Population 12.3
Tropheus "Kirschfleck", T. "Cherry Spot", Kirschfleck Moorii
Bulu Pnt, Tanzania, W, B, 10-12 cm

photo: P. Schupke

A97372-2 *Tropheus* sp. Bulu Point; Population 12.3
Tropheus "Kirschfleck", T. "Cherry Spot", Kirschfleck Moorii
Bulu Pnt, Tanzania, W, B, 10-12 cm

photo: P. Schupke

A97372-3 *Tropheus* sp. Bulu Point; Population 12.3
Tropheus "Kirschfleck", T. "Cherry Spot", Kirschfleck Moorii
Bulu Pnt, Tanzania, W, B, 10-12 cm

photo: P. Schupke

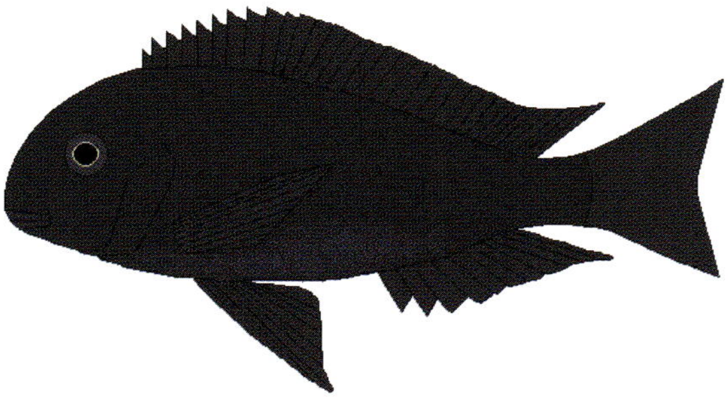

A97373 *Tropheus* sp. Kabezi, Linie 13, Population 13.0

Near Kabezi village, Bureundi, W, 10-12 cm

drawing: P. Schupke

A97374-4 *Tropheus* sp. "Red Belly"

Bulu Point, Tanzania, W, B, 10-12 cm

photo: K. A. Ackerbauer

A97374-4 *Tropheus* sp. "Red Belly"

Bulu Point, Tanzania, W, B, 10-12 cm

photo: K. A. Ackerbauer

A97374-4 *Tropheus* sp. "Red Belly"

Bulu Point, Tanzania, W, B, 10-12 cm

photo: K. A. Ackerbauer

A97374-4 *Tropheus* sp. "Red Belly"

Bulu Point, Tanzania, W, B, 10-12 cm

photo: K. A. Ackerbauer

A97374-4 *Tropheus* sp. "Red Belly"

Bulu Point, Tanzania, W, B, 10-12 cm

photo: K. A. Ackerbauer

A97375-4 *Tropheus* sp. "Red Belly"

Sibwesa, Tanzania, W, 10-12 cm

photo: P. Schupke

A97375-4 *Tropheus* sp. "Red Belly"

Sibwesa, Tanzania, W, 10-12 cm

photo: P. Schupke

A97376-4 *Tropheus* sp. "Karilani Red"

Karilani island, Tanzania, W, 10-12 cm

photo: P. Schupke

A97376-4 *Tropheus* sp. "Karilani Red"

Karilani island, Tanzania, W, 10-12 cm

photo: P. Schupke

A97376-4 *Tropheus* sp. "Karilani Red"

Karilani island, Tanzania, W, 10-12 cm

photo: P. Schupke

A97376-4 *Tropheus* sp. "Karilani Red"

Karilani island, Tanzania, W, 10-12 cm

photo: P. Schupke

A97377-4 *Tropheus* sp. "New Kirsch"

Origin unknown, W, 10-12 cm

photo: P. Schupke

A97377-4 *Tropheus* sp. "New Kirsch"

Origin unknown, W, 10-12 cm

photo: P. Schupke

A97377-4 *Tropheus* sp. "New Kirsch"

Origin unknown, W, 10-12 cm

photo: P. Schupke

A97941-4 *Tropheus* sp. Kalambo; Population 8.12, Mutant
Tropheus "Kalambo", Lemon Spot Tropheus, Golden Kalambo
South of Kalambo river to Chituta bay, Zambia, W, B, 10-12 cm

photo: B. Bender

A97941-4 *Tropheus* sp. Kalambo; Population 8.12, Mutant
Tropheus "Kalambo", Lemon Spot Tropheus, Golden Kalambo
South of Kalambo river to Chituta bay, Zambia, W, B, 10-12 cm

photo: B. Bender

A97941-4 *Tropheus* sp. Kalambo; Population 8.12, Mutant
Tropheus "Kalambo", Lemon Spot Tropheus, Golden Kalambo
South of Kalambo river to Chituta bay, Zambia, W, B, 10-12 cm

photo: B. Bender

In jeder

Aqualog - *news*

internationale Zeitung für Aquarianer

erscheinen als **stickups** Ergänzungen
mit neuen Fischbildern, die Sie

hier

einkleben sollten, damit Sie immer "up-to-date" sind.

In every

Aqualog - *news*

international newspaper for aquarists

you will find **stickups** as supplements
which should be sticked in

here

to make sure that you are always up-to-date

Ihr Aqualog-Nachschlagewerk
Your Aqualog reference work

Vervollständigen Sie Ihr Nachschlagewerk durch weitere Bücher der Aqualog-Reihe:

Complete this reference work with further volumes of the Aqualog series:

ISBN 3-931702-13-8

ISBN 3-931702-04-9

ISBN 3-931702-07-3

ISBN 3-931702-10-3

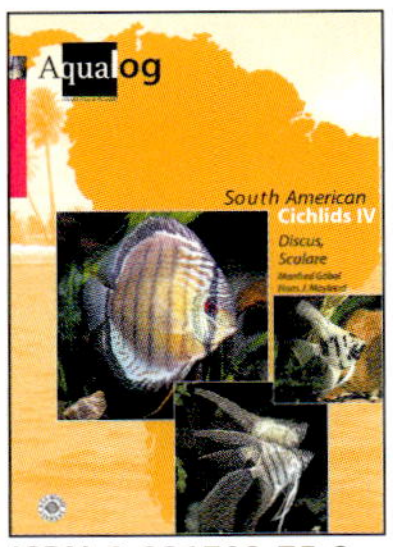

ISBN 3-931702-75-8

ISBN 3-931702-79-0

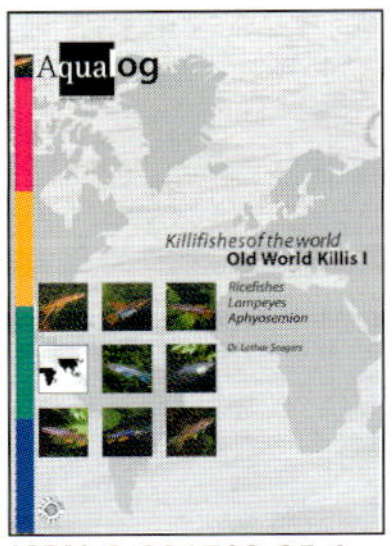

ISBN 3-931702-25-1

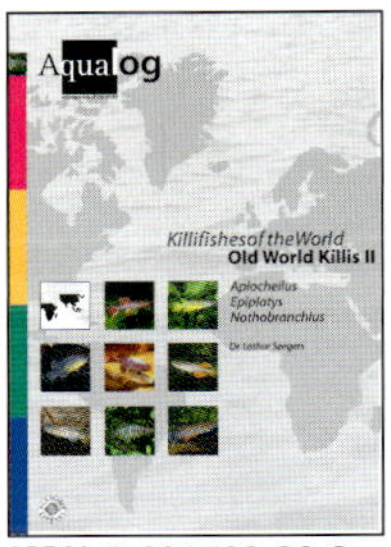

ISBN 3-931702-30-8

ISBN 3-931702-76-6

ISBN 3-931702-21-9

ISBN 3-931702-77-4

ISBN 3-931702-01-4

Mehr Informationen direkt bei

For more information please contact

Aqualog Verlag
Liebigstraße 1, D-63110 Rodgau/Germany
Fax: +49 (0) 61 06 - 64 46 92,
email: acs@aqualog.de
Internet: http://www.aqualog.de

ISBN 3-931702-78-2

ISBN 3-931702-93-6

ISBN 3-931702-80-4

Alle Aqualog-Produkte erhalten Sie im Zoofachhandel und überall auf der ganzen Welt. Wir nennen Ihnen gerne Bezugsquellen.

You can obtain all Aqualog products everywhere in the world. Contact us for addresses.

ISBN 3-931702-61-8 (deutsche Fassung)

ISBN 3-931702-60-X *(English edition)*

Aqualog*news*

Informiert top aktuell zu folgenden Themen:

Up-to-date information on following topics:

- Süß- und Seewasserfische / *marine and freshwater fishes*
- Terraristik / *terraristic*
- Neuentdeckungen / *new discovered fishes*
- Aquarienpflanzen / *plants for the aquarium*

Fordern Sie Ihr Gratis-Exemplar der englischen oder deutschen Ausgabe beim Verlag an!

Order your free specimen copy of the German or English edition at the publisher!

VERDUIJN CICHLIDS
Specialized in cichlids from all over the world
wholesale & retail of wild and tank-raised fish
Over 30 years of experience!
Wollefoppenweg 107
2761DL Zevenhuizen ZH The Netherlands
tel+31(0)104550253 fax+31(0)104566738
www.verduijncichlids.com

FOR YOUR CICHLIDS
AND OTHER FISH...
Super Spirulina Forte is one of the most valuable, scientifically-based flake foods with very high (36%!) content of the purest Spirulina platensis algae available for use in professional aquarium fish breeding.
Super Spirulina Forte is especially recommended for African cichlids, which feed on algae and all other fish with high nutritional needs.
Super Spirulina Forte is already available. Ask for it in the aquarium shops.
Tropical
SUPER SPIRULINA FORTE
HIGH QUALITY PRODUCT
SCIENTIFICALLY BASED DIET
36 %
1200 ml
Since 1977
Tropical®
ANDRZEJ OGRODNIK
PL 41-507 Chorzów, ul. Opolska 25
tel. (+4832) 249 92 10, fax (+4832) 249 94 58
www.tropical.com.pl
PROFESSIONAL AQUARIUM
© - TROPICAL A. OGRODNIK 2003

THE ORIGINAL O.S.I®
SPIRULINA FLAKE
Spirulina Aquarium Flake Food
European Distributor
Smulders Wholesale BV
www.smulders.nl

Aqua Dream Aquarien
swiss design
Leidenschaft und Perfektion
Amazonas II
swiss system
Internal Filters System for the future
Aqua Dream Design GmbH
Aquaristikbedarf Engros international
Schrofenstrasse 4/6
CH-8580 Amriswil
Tel./Fax (0041) 071/411'76'37
e-mail; info@aquadream.ch
www.aquadream.ch

Cichliden-Spezialisten weltweit
Cichlid Specialists Worldwide

Ostafrikanische Cichliden
K. Ruffert
Herkentruper Str. 23
D-48329 Havixbeck
Tel: 02507- 571425
Email cichlid-power@t-online.de
www.cichlid-power.de

Cichlide Centret
Baekkeskovvej 9
DK-2665 Vallensback Strand
Denmark
Tel: 0045 (-43) -530016
cichlidecentret@cichlidecentret.dk
www.cichlidecentret.com

Afrika Schuppen
Angelika Prüße
Südstrasse 22
D-38170 Dahlum
Tel: 05332 2380
Email fischzucht-pruesse@t-online.de
www.tanga-mala.de

Aqua Planning
Nourneystraße 5-7
D-40822 Mettmann
Deutschland
Tel: 02104 16539

Aquafarm
Westfalenring 6
D-45739 Oer-Erkenschwick
Deutschland
Tel: +49 (0) 2368 692724
Email info@aquafarm.de
www.aquafarm.de

Malawi Tanganjika Shop
Niedermöllern 29
D-06628 Möllern
Deutschland
Tel./Fax: 034463 26346
malawi-tanganjika-shop@t-online.de
www.malawi-tangajika-shop.de

Verduijn Cichlids
Wollefoppenweg 107
2791DL Zevenhuizen
Nederland
0031 (10) 4550253
Email info@verduijncichlids.com
www.verduijncichlids.com

Canadian Rift Lake Cichlid Assoc, Inc.
1166 Gordon Str. Guelph,
Ontario N141H2 Canada
Tel. +1 519 8226130
Fax +1 519 8368899
Email crlca@rogers.com
Website: http://www.crlca.com

Halten Sie Ihr Aqualog-Lexikon über Jahre aktuell
Keep your Aqualog-Lexicon up-to-date for years

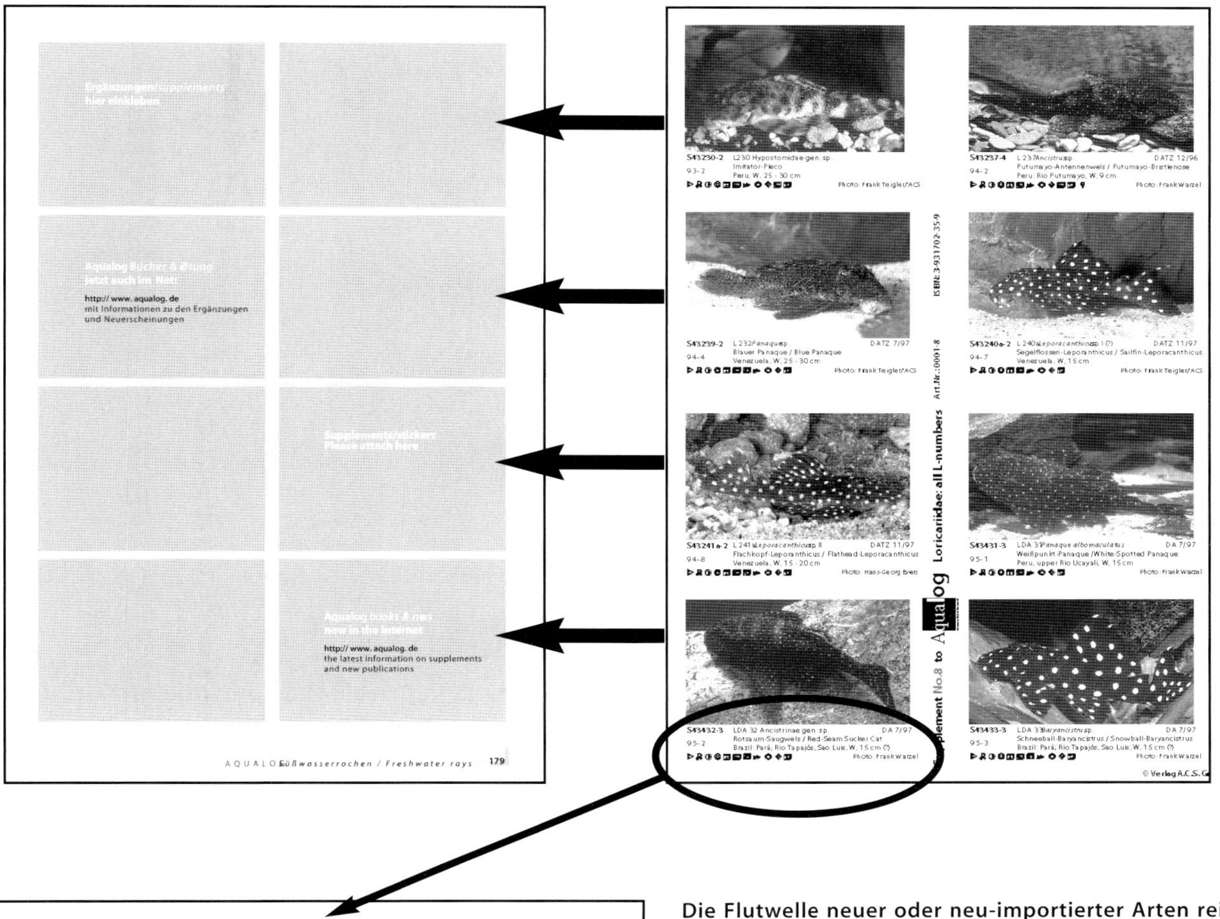

① **S43432-3** LDA 32 Ancistrinae gen. sp. DA 7/97
② 170/95-2 Rotsaum-Saugwels / Red-Seam Sucker Cat
Brazil: Pará; Rio Tapajós, Sao Luis, W, 15 cm (?)
③ Foto: Frank Warzel ④

① **Code Nummer**
Code number

② **1. Zahl**: fortlaufende Bildnummer
***1. number**: continuous picture number*

2. Zahl: Seitennummer im Buch
***2. number**: page number in the book*

3. Zahl: Bildnummer auf der Seite (durchlaufend numeriert von 1–8 von oben links nach unten rechts)
***3. number:** picture number on the page (continuously numbered from 1–8 from the top left corner to bottom right)*

③ **Symbol Leiste**
Aqualog-**Bücher**
Symbol text
(Aqualog-books)

④ **Bildautor**
Photographer

Die Flutwelle neuer oder neu-importierter Arten reißt nicht ab. Daher haben wir uns entschlossen, Ergänzungsbögen mit je acht Einklebebildern zu einem Buch zu erstellen. Lieferbar über den guten Zoofachhandel und den Buchhandel zum Preis von DM 4,80 pro Stück. Viel Freude damit!

The flood of new or newly-imported species doesn´t stop. So we have decided to print supplements with eight stickers each (each supplement contains pictures for a single volume of AQUALOG). They can be ordered at well-equipped pet shops or in any book shop. We hope you enjoy them!

Bitte beachten Sie nebenstehendes Schema, bevor Sie die Bilder einkleben. Die Ergänzungen erscheinen nicht zwangsläufig in der Reihenfolge, in der sie eingeklebt werden, sondern in der Reihenfolge ihrer Verfügbarkeit. Wenn wir z. B. anfangs nur das Bild eines Weibchens als Ergänzung haben, jedoch sicher sind, früher oder später auch das Bild eines Männchens zu bekommen, sollte das Bildkästchen links vom Weibchenbild frei bleiben.

Please follow the scheme given here, before you stick in the pictures. The supplements are not necessarily in the correct order. For example: if we have only the photo of a female, but we are sure to get the photo of the male sooner or later, too, please keep the space to the left of the female free.

Ergänzungen/*supplements*
hier einkleben

Aqualog *Bücher & Zeitung*
jetzt auch im Net:

http:// www. aqualog. de
mit Informationen zu den Ergänzungen
und Neuerscheinungen

Supplements/stickers
Please attach here

Aqualog *books & news*
now in the Internet

http:// www. aqualog. de
the latest information on supplements
and new publications

Ergänzungen/*supplements*
hier einkleben

Aqualog *Bücher & Zeitung*
jetzt auch im Net:

http:// www. aqualog. de
mit Informationen zu den Ergänzungen
und Neuerscheinungen

Supplements/stickers
Please attach here

Aqualog *books & news*
now in the Internet

http:// www. aqualog. de
the latest information on supplements
and new publications

Ergänzungen/*supplements*
hier einkleben

Aqualog *Bücher & Zeitung*
jetzt auch im Net:

http:// www. aqualog. de
mit Informationen zu den Ergänzungen
und Neuerscheinungen

Supplements/stickers
Please attach here

Aqualog *books & news*
now in the Internet

http:// www. aqualog. de
the latest information on supplements
and new publications

Ergänzungen/*supplements*
hier einkleben

Aqualog *Bücher & Zeitung*
jetzt auch im Net:

http:// www. aqualog. de
mit Informationen zu den Ergänzungen
und Neuerscheinungen

Supplements/stickers
Please attach here

Aqualog *books & news*
now in the Internet

http:// www. aqualog. de
the latest information on supplements
and new publications

Ergänzungen/*supplements*
hier einkleben

Aqualog *Bücher & Zeitung*
jetzt auch im Net:

http:// www. aqualog. de
mit Informationen zu den Ergänzungen
und Neuerscheinungen

Supplements/stickers
Please attach here

Aqualog *books & news*
now in the Internet

http:// www. aqualog. de
the latest information on supplements
and new publications

Ergänzungen/*supplements*
hier einkleben

Aqualog *Bücher & Zeitung*
jetzt auch im Net:

http:// www. aqualog. de
mit Informationen zu den Ergänzungen und Neuerscheinungen

Supplements/stickers
Please attach here

Aqualog *books & news*
now in the Internet

http:// www. aqualog. de
the latest information on supplements and new publications

Ergänzungen/*supplements*
hier einkleben

Aqualog *Bücher & Zeitung*
jetzt auch im Net:

http:// www. aqualog. de
mit Informationen zu den Ergänzungen
und Neuerscheinungen

Supplements/stickers
Please attach here

Aqualog *books & news*
now in the Internet

http:// www. aqualog. de
the latest information on supplements
and new publications

Ergänzungen/*supplements*
hier einkleben

Aqualog *Bücher & Zeitung*
jetzt auch im Net:

http:// www. aqualog. de
mit Informationen zu den Ergänzungen
und Neuerscheinungen

Supplements/stickers
Please attach here

Aqualog *books & news*
now in the Internet

http:// www. aqualog. de
the latest information on supplements
and new publications

Ergänzungen/*supplements*
hier einkleben

Aqualog *Bücher & Zeitung*
jetzt auch im Net:

http:// www. aqualog. de
mit Informationen zu den Ergänzungen
und Neuerscheinungen

Supplements/stickers
Please attach here

Aqualog *books & news*
now in the Internet

http:// www. aqualog. de
the latest information on supplements
and new publications

Ergänzungen/*supplements*
hier einkleben

Aqualog *Bücher & Zeitung*
jetzt auch im Net:

http:// www. aqualog. de
mit Informationen zu den Ergänzungen
und Neuerscheinungen

Supplements/stickers
Please attach here

Aqualog *books & news*
now in the Internet

http:// www. aqualog. de
the latest information on supplements
and new publications

Index , Codenummern / codenumbers

Literaturverzeichnis / Literature cited

BARIC, S. (2000): Impact of Lake Level Fluctuations on Population Structure and Speciation in Rock Dwelling Cichlids from Lake Tanganjika. Dissertation, Insbruck.

BAUER, R.(1991): Tierärztliche Heimtierpraxis; Erkrankungen der Aquarienfische; Paul Parey Verlag

BRICHARD, P. (1978): Fishes of Lake Tanganjika; T.F.H. Publications, Neptune City

BRICHARD, P. (1997): Atlas der Tanganjikacichliden, Band 1; Bede Verlag, Ruhmannsfelden

DATZ- Sonderheft (1998): Tanganjikasee; Verlag Eugen Ulmer, Stuttgart

HERRMANN, H. J. (1990): Die Buntbarsche der Alten Welt, Tanganjikasee, Verlag Eugen Ulmer, Stuttgart

HERRMANN, H. J. (1996): Aqua Lex Katalog, Tanganjikasee; Dähne Verlag, Ettlingen

HERRMANN, H. J. (1997): Tropheus; Dähne Verlag, Ettlingen

HÜCKSTEDT, G. (1973): Aquarienchemie; Kosmos-Verlag, Stuttgart

KONINGS, A. & H. W. Dieckhoff (1992): Geheimnisse des Tanganjikasees, Cichlid Press, El Paso

KONINGS, A. (1988): Tanganjikacichliden; Verduijn Cichlids, Zevenhuizen

KONINGS, A. (1992): Das Cichlidenjahrbuch; Band 2; Cichlid Press, El Paso

KONINGS, A. (1993): Cichliden, artgerecht gepflegt; Cichlid Press, El Paso

KONINGS, A. (1994): Das Cichlidenjahrbuch; Band 4; Cichlid Press, El Paso

KONINGS, A. (1995): Das Cichlidenjahrbuch; Band 5; Cichlid Press, El Paso

KONINGS, A. (1998): Tanganjikacichliden in ihrem natürlichen Lebensraum; Cichlid Press, El Paso

KRAUSE, H. J. (1995): Handbuch Aquarienwasser; Bede Verlag, Ruhmannsfelden

MAYLAND, H. J. (1978): Cichliden und Fischzucht, Aquarienpraxis III, Landbuch Verlag, Hannover

NEERGAARD, S. (1982): Tanganjikacichliden; Kernen Verlag, Stuttgart

REICHENBACH- KLINKE, H. H. (1968): Krankheiten der Aquarienfische. Alfred Kernen Verlag Stuttgart

SCHNEIDEWIND, F. (1996): Das große Buch der Tropheus-Cichliden. Bede Verlag, Ruhmannsfelden

SELZLE, H. (1987): Technische Aquaristik; Eigenverlag?

STAECK, W. & H. Linke (1998): Afrikanische Cichliden II: Buntbarsche aus Ostafrika. Tetra Verlag, Münster

STAECK, W. (1975): Cichliden, Verbreitung, Verhalten, Arten; Engelbert Pfriem Verlag, Wuppertal, Elberfeld

STAECK, W. (1975): Die südlichen Rassen von Tropheus moorii; Aquarienmagazin, Heft 12

STAECK, W. (1977): Cichliden, Verbreitung, Verhalten, Arten; Band 2 (Supplement); Engelbert Pfriem Verlag, Wuppertal, Elberfeld